# Life Resonance

## How to Enrich Your Life

By Steve Preston

3rd Edition

© Copyright 2012, Steve Preston
All rights reserved.
No part of this book may be reproduced, stored in a retrieval system, or transmitted by any means, electronic, mechanical, photocopying, recording, or otherwise, without written permission from the author.

# Table of Contents

LIFE RESONANCE ............................................................................................. 1

TABLE OF CONTENTS ....................................................................................... 3

WHAT'S IN THE BOOK? .................................................................................... 6

RELATIVITY ..................................................................................................... 24

10-DIMENSIONS .............................................................................................. 30

STRUCTURAL GROUP ..................................................................................... 34

OPERATIONAL GROUP ................................................................................... 37

ETHEREAL GROUP .......................................................................................... 40

MUTUAL PERPENDICULARITY ....................................................................... 46

THREE SECRETS OF TIME ............................................................................... 49

SPACE RESONANCE ........................................................................................ 52

QUALITY OF RESONANCE .............................................................................. 54

LIFE RESONANCE ........................................................................................... 58

SELF ACTUALIZATION .................................................................................... 59

LIFE RESONANCE LEVELING .......................................................................... 65

FAITH .............................................................................................................. 68

ATOMIC FUSION ............................................................................................. 72

VIBRATIONAL ALCHEMY ................................................................................ 77

| | |
|---|---|
| COMMON MATERIAL FREQUENCIES | 83 |
| BEAT FREQUENCIES | 87 |
| VIBRATING IN HISTORY | 90 |
| VIBRATION STAFF | 92 |
| MOSES STAFF | 97 |
| LEVITATION AND THE DJED | 102 |
| ARC POWER ANOMALY | 106 |
| BOOK OF SECRETS | 110 |
| PEOPLE IN GROUPS | 115 |
| TESLA VIBRATES | 117 |
| LEVITATION | 123 |
| HUMAN COMBUSTION | 126 |
| ED LEEDSKALNIN | 134 |
| JOHN HUTCHISON | 137 |
| INVISIBILITY AND ALCHEMY | 140 |
| BEYOND | 143 |
| SOUL AND SPIRIT | 145 |
| LIGHT | 147 |
| OUT OF BODY CONSCIOUSNESS | 149 |
| CARBON 12 ANTHROPICS | 151 |
| THE ANTHROPIC PRINCIPLE | 152 |
| JEWS | 158 |

| | |
|---|---|
| CHRISTIANS | 163 |
| AFTER DEATH | 166 |
| CONSERVATION OF EVERYTHING | 169 |
| SOUL JOURNEY | 172 |
| RE-ENTRY OF LIFE | 174 |
| LIFE AND DEATH? | 178 |
| DISSIPATING LIFE | 182 |
| HOW MANY HEAVENS? | 185 |
| DREAMING | 186 |
| WHY CONTINUOUS LIFE | 188 |
| LIFE IN DEATH | 191 |
| GOD AND THE TRINITY | 193 |
| FINAL END NEW BEGINNING | 196 |
| CONCLUSIONS | 198 |
| ABOUT THE AUTHOR | 203 |

# What's in the Book?

Have you ever heard the term "Think and Grow Rich"? How about "The Power of Positive Thinking" or Thomas Maslow's "Self Actualization"? These and many other catch phrases are simply stating what Einstein and his cronies determined after years of struggling with thousands of blackboards filled with mathematics. Our lives can change its resonance to assure modification in reality. Today there is an entire science designed around this powerful capability inside each of us. It is called Anthropic science. As our resonance frequency increases, our control over our environment increases.

*Somehow, people affect the universe. Their consciousness changes what we call Reality.*

This is not as simple as it sounds in that we don't have a knob to turn or anything like that. Because of the nature of this subject, I must tell you, a substantial amount of this book will be difficult to absorb, initially. Please do not put it down when something seems odd to you. The very nature of controlling what we call reality is so very odd that we will have to relearn living to some extent. What we will find is that this power to control our perceived reality is associated with something we can call

<u>vibrational resonance</u>. This is sort of like "being in tune with the universe". If a person understood and was willing to be in tune with the universe, many elements of the reality sensed by him could be controlled.

God incarnate [Jesus] said it this way. *"With faith as tiny as a grain of mustard-seed, one could move mountains"*.

I know it sounds physically impossible or like some esoteric conditional thing whereby God changes physics if you have faith in him, but that its not what it says at all.

It is saying if you believe enough----YOU can modify what we call reality. You typically can't because <u>you limit yourself</u> to what you believe to be reality rather than what IS reality and what we are. If we are to truly understand ourselves and understand life, we must first start with this basic thing. By the way the <u>"faith" described by God incarnate, was not a belief that God was God</u>, it was an understanding and removal of your consciousness from Carnal thinking to something could allow our **"life resonance"** frequency to increase--- to become more **"holy"** so to speak. In this book I will show you why this is true, the science behind it, and some thoughts on how you can modify reality for short periods of time.

*We can affect reality! How much we affect it is dependent on something we can call life resonance. Life resonance is not specifically religious, but it is separation from belief in this "apparent reality".*

The concept of "meditation" [even meditation without God centric view] is a very minor focus on changing our reality perception. This Life Resonance is not only important while we are alive, but <u>also in how we react in death</u>. If our consciousness is reality; in "this" reality, our consciousness cannot cease to exist. This whole concept seems so important, I thought this book would help people understand how to model it, how to change it, how to live with it, and how to understand that death is not and cannot be the end. We first start with what we can loosely call our consciousness. It is our consciousness and our spirit that affect reality. Freud thought of this as the "ID" consciousness. If you are not sure what spirit means, I'll get into that a little later.

Because our consciousness and our spirit both affect reality, both must be defined as dimensions of this reality. I'm sure you are already thinking this is going to be some esoteric, goobly-gooky challenge to sanity, but hopefully, I can explain it in a way to dispel those feelings. In this book, I am going to describe to you what we know about the most mysterious dimensions of the Universe. They are not length or height or width as you were told in school and neither is the quasi-dimension of time.

### Elimination of 3-Dimensions

Before we get into the almost immeasurable characteristic of Consciousness let me first start with some easier things. I promise to help you understand how to affect the universe better by the end of the book so stay with me.

*We live in a 3- dimensional world---WRONG!*

Everyone you meet, everyone you see, everyone you ask tells you the same thing. Even college professors tell you about a 3-dimensional world where matter is composed of these three components—length, height, width, or something we call volume. We could say we live in a volum-ated universe. It's so obvious, no one would even begin to question such a thing—--------if you listen to these characters. Volume not only characterizes what we can touch, fell, see, but also it supposedly allows for the space of nothingness between the various somethings. Wait a minute!!!! You can't have a volume of nothing. Some talk about a volume of vacuum, but that is truly a misnomer. Our huge volum-ated universe has only a tiny amount of volum-atable mass and almost all the rest SEEMS to be nothing at all. Today we know that the "nothing stuff" is just as important as the things we call mass.

## What About Time?

Before we go on, we had better look the quasi-dimension of time. Things just couldn't happen if everything was just mass, so someone came along and bellowed out that time must be a dimension so things could happen in sequence. If time disappeared, everything would happen simultaneously, and that would be no good for us or anyone else. Einstein came along and said that if you go the speed of light, no time passes for you, He went on to say that if everyone went the speed of light, there would be no time period. Everyone just stood there. "We have to have time as a dimension", they demanded, so going the speed of light was sort of outlawed. This idea that if everyone went the speed of light would stop time made it sound like people could control reality. Some poked pins in pictures of Einstein, some scoffed, some laughed, but many started testing his theories and found that consciousness DOES control reality and time is sort of an offset of consciousness.

## Created Matter

The height, length, width was adopted, but many, many things did not fit; matter spontaneously erupting at the edge of a black hole, for instance. At this "Event Horizon", as Steve Hawkins explained, spontaneous mass could be realized. To allow it to be recognized in our volume universe, the term "black matter" was pushed into our classrooms. It was matter that wasn't really there, but could change at one of these black hole edges. Oops! I said black hole not realizing they can't exist in a 3 dimensional world either. I'm sorry for the slip, but it was already down on paper before I knew it.

## Religion

Oops! I did another boo-boo starting to talk about religious characteristics. People began to say things like, *"It is physically impossible for Angels to miraculously appear, spirits can't exist, the Holy Spirit must be excused from reality, and walking on water must have been a hoax."* Rather than trying to explain how religion can be real, most people, even those who claimed to be religious, began separating religion from reality. On Sunday, the "religious world" was thought about but on Monday, people went back to a stagnate universe with "3-dimensions plus time" and kept the 2 completely separate so they would have no conflict. Soon, Einstein got back in the picture so that people could believe in God again and still have science. He didn't start out to make it easier to have angels, and the Holy Spirit and all the rest. He started a transformation with the Sun.

## The Sun

Besides trying to find out how religion worked, we still had problems finding out why everything else worked. Why wasn't everything made up of hydrogen as the law of Entropy tells us everything goes towards its most disorganized state and there is simply no way to make large atoms.

*Einstein came along and said $E=MC^2$ and everyone cheered. "Mass is really energy rather than substance". Mass doesn't exist except as an energy.*

Nobody really knows what this ENERGY stuff was, so it whole mass is not real didn't sink in for a while. If a lot of mass gets together, or should I say energy, fusion makes big atoms. A bunch of hydrogen "energy" got together in a clump and started making something we call gravity, more hydrogen was sucked in and more gravity was made until WHAM! The center of the hydrogen ball began making new atoms and in the process, it released these photon things. Depending on the way it vibrated, it appeared to make colored light. Hydrogen normally made yellowish energy bundles we called photons, but getting the photons to the surface of this massive ball of gas added energy and made it vibrate differently. Why that mattered, no one really knew, but colors were made just the same. The main things to know here are that there is a lot of vibrating going on and mass has no real mass, according to Einstein and most others trying to satisfy the mathematics of the universe. Without mass being mass, the 3 dimensional definition of our universe was in shambles so they tried to use photons to allow height, length and width to define our universe.

## Photon

We started looking a light and the scientific powers announced that light was sometimes a particle and sometimes it was something called a wave.

While that is a whole lot of malarkey, let's look at this wave thing. What these guys said was that light sometimes doesn't exist as a thing. Instead, a whole bunch of nothingness was vibrated faster and faster until people saw stuff. Still faster the nothing vibrated and people could see through people. If these photon things still vibrated faster they, collectively, are called "cosmic rays" which started killing everything in "their" path. OOPS! I can't identify cosmic rays as a thing. "They" are defined as nothingness vibrating which can't be in a 3-dimensional world.

*Take a minute for all this stuff to sink in. When you are ready, continue!!*

Below is a chart of the thing we call light.

| Description | Cycle/sec | Frequency | Effect |
|---|---|---|---|
| Helpful Infrared light | 1 x10-6 | 30 x 1013 | Invisible thing |
| Visible light | 4 x10-7 | 75 x 1013 | Visible thing |
| Dangerous X-rays | 1 x10-8 | 30 x 1015 | Penetrates bone |
| Deadly Gamma Rays | 1 x10-9 | 30 x 1016 | Invisible thing that destroys |

I know we still have that "sometimes it is a particle" thing to fall back on. Oh well! People, SOMEHOW, accepted the definition and scientists were happy. After all, we could see light---right?

### Electro-Magnetics

Someone said, "What if we slow this light stuff down; can we make radios work?" Wow! There it was, slow vibrating light was making radios work and the study of electro-magnetics brought us AM, FM, Televisions and the internet all without any of these slow photons getting MASS so that they could be defined in a 3-dimensional world. Below is an expanded "light" chart.

| Name | Maximum Wavelength [m] | Highest Freq. [Hz] |
|---|---|---|
| Pure DC Voltage/ electric potential | $5 \times 10^{12}$ | ~0 |
| Human hearing* | $1 \times 10^{4}$ | $20 \times 10^{3}$ |
| VHF [radio] | $1 \times 10^{0}$ | $30 \times 10^{7}$ |
| UHF [radio] | $1 \times 10^{-1}$ | $30 \times 10^{8}$ |
| SHF [radio] | $1 \times 10^{-2}$ | $30 \times 10^{9}$ |
| EHF [radio] | $1 \times 10^{-3}$ | $30 \times 10^{10}$ |
| Microwaves | $2.5 \times 10^{-4}$ | $12 \times 10^{12}$ |
| Infrared [light] | $1 \times 10^{-6}$ | $30 \times 10^{13}$ |
| Visible light | $4 \times 10^{-7}$ | $75 \times 10^{13}$ |
| X-rays | $1 \times 10^{-8}$ | $30 \times 10^{15}$ |
| Gamma Rays | $1 \times 10^{-9}$ | $30 \times 10^{16}$ |
| Pure Magnetism | | Really fast |

*Note on the chart- The human hearing frequencies are not normally thought of as electromagnetic, but what I'm talking about hear is the feelings one gets from certain sounds. Lower frequency sound puts a darkness in our images and faster vibrations bring in more light so there is some type of connection.

## String Theory Scientists

People started going crazy and one of them said let's get more dimensions. Another said, "We can't have more than 3 dimensions. What would it look like?" Others simply said, *"We'll make each dimension separately defined as a wiggling string just going around and if a length, height, width dimension got together, matter would be Apparent."* After some math, it was determined that there must be at least <u>10 dimensions and maybe even 11</u>. That is as far as many wanted to go so they backed away from their findings by saying, *"The way we can still have 3-dimensional volume is that most of the dimensions were compactified".*

Sure, it was a made up word, but it allowed matter to be made in a <u>"3ish dimensional universe"</u>. I know the whole compactified [meaning zeroed out or not used] dimensions don't make sense, but the math said the 10 dimensions were there. String theorists simply made a new word to allow us to use our length, height, width and be comfortable.

## Einstein Again

**Einstein** started looking at atoms and particles inside atoms. *He said that there was a unified particle that makes up everything.* He knew it wasn't an atom, because there were over 100 different types of them, BOSONs like electrons, were out as they found that there were many sub-particles making up them. It wasn't even Quarks as many types of these almost particles were found. Something even more primary made up everything. It only made sense so everyone started looking for it. The electron was broken down from quarks, to something called fermions. These fermion things didn't have mass but they had things like gravitational pull or nuclear attraction without really being there. Unfortunately, we found all types of these things that generally had no height, length, or width, but still existed. Following are some of these not quite particles.

| Fermion Name | Information about its Oddness |
|---|---|
| Neutrino | This is a quasi- particle that is a component of a Up-quark, Three types known [Electron-neutrino, Muon-neutrino, and Tau-neutrino.] They have almost no reaction with matter and can pass through the Earth--- They have no apparent mass. |
| Electron hole | A lack of electron in a valence band. While everyone uses these electron holes, to interpret electricity, they don't completely exist. |
| Photon | This previously was considered a Boson that has no apparent mass but has electromagnetic properties. It can be "modeled" with 2 quarks or equivalent particles. It exhibits no Gravitational force, but instead it SOMEHOW, makes light. Like many of these vibrational components, the vibrational travel is faster than the speed of light. |
| Graviton | Like the photon, this is another fermion possibility that has no apparent mass, yet it exhibits a strong gravitational force. This suggests that an even quantity of quarks is combined in its makeup. |
| Chargon | A quasi-particle produced from an electron spin-charge separation |

| | |
|---|---|
| Config-uron | An excitation in an amorphous material associated with breaking of a chemical bond |
| Exciton | A bound state of an electron and a hole |
| Fracton | A collective quantized vibration on a substrate with a fractal structure. |
| Holon | A quasi-particle resulting as a result of electron spin-charge separation |
| Libron | A quasi-particle associated with the motion of molecules in a crystal |
| Magnon | A coherent excitation of electron spins in a material |
| Majorana fermion | A quasi-particle equal to its own antiparticle in superconductors |
| Phason | Vibrational modes in a quasi-crystal from atomic rearrangements |
| Phonon | Vibrational modes in a crystal lattice associated with atomic shifts |
| Plasmon | A coherent excitation of a plasma |
| Polaron | A charged quasi-particle that is surrounded by ions in a material |
| Polariton | A mixture of photon with other quasi-particles |
| Roton | Elementary excitation in super-fluid Helium-4 |
| Soliton | A self-reinforcing solitary excitation wave |
| Spinon | A quasi-particle produced as a result of electron spin-charge separation |

| | |
|---|---|
| Gluon | A quasi-particle that can cause an interaction between Mesons and quarks opposite to gravity |
| Lepton | A quasi-particle [no-mass] that does not attract nuclear force |

"How in the world are we going to use the 3-dimensions for defining the universe?" Many were silent but the string dimensionalists compactified away and said, "I told you so!" Other scientists started thinking about all the dimension strings and rationalized that there must be other universes using some of these dimensions. Maybe the universe of "Heaven" wasn't just a myth.

## Parallel Universes and Heaven

Extra dimensions worked by allowing integrations of many universes that coexist with our 3ish dimensional one. Scientist said, "We had better get some more universes. We won't see them, but people have already thought about Heaven and some even addressed a "Purgatory" place, as unseen universes, so it will be an easy transition." Ten-dimensions turned into 10 interlinked universes. Things could travel from one to another in some way and a new explanation rose up called "super symmetry".

## Super-Symmetry

In essence, this said that if anything came into existence in our universe, something disappeared in an adjacent universe. If something gets larger here, something gets smaller in another universe. Angels could come from

Heaven to here provided, that something went from here to there that was the same "mass". Massless things could be accounted for because they would have huge masses in an adjoining universe. On and on we could go to show why this was a helpful definition of existence.

*It almost "de-religious-ized" religion.*

--EXCEPT!!!-- You still had to start from somewhere and that took a creator and all these explained dimensions still didn't really help with all the stuff locked in our own universe like this Big Bang Thing.

### Einstein Again

He and others said the universe started with this huge explosion and everything was expanding away from the blast ever since. Several problems occurred, but many were explained away with the linked Heaven universe. Others were not explained away. One is the center anomaly. The center of the universe was located by something called red shift and some math and it was determined to be approximately, where the Earth is. If that was the center of the blast, the earth would have been exploded as it was messing up the 3-dimensional world concept.

People ignored the problem and Einstein started investigating a new one. He determined that as the energy of the universe reaches its limits, it would eventually leave our universe or the energy levels at the center of the universe would simply get less and less and less until everything was dead. There would be no mass left, not light left, no energy left.

Einstein worried with trying to find the tiny quasi-particle that made up everything that made the universe and how energy SOMEHOW gets replenished in the center of the universe for many years and then he died without knowing the truth.

### Traveling Through Nothing

Before he died, he started to think about getting around in space and going through all of the "EMPTY AREAS". If light needed to vibrate to be light, it had no way of doing that through a nothing, so the concept of Aether took shape. Aether would be something that has no mass, but existed to allow the fermion things and everything else to be suspended in. These Aether things did not restrict motion, so Einstein called them "Squishy Things". I know Aether doesn't fit in 3 dimensions, but let's see what was said about this stuff.

### What is Aether?

Here is sort of a secret illustration that I use to think about this Aether stuff. Noise cancellation speakers take 2 identical sounds and invert one of them so that the vibrations are exactly opposite. When the 2 sounds are put together, they disappear. If ½ of the fermion quasi particles vibrated out of phase with another half, they cannot make a visible object as their VIBRATION is cancelled. They are still there, and in some way allow transmissions through. They simple make the "vibrational element" neutral. This is what Einstein called "Aether". Let me make this perfectly clear to you so we can determine what the universe is. Aether is cancellation of the vibrational characteristics of 2 no-mass things.

### *My head Hurts!!*

Sounds like we are really trying to force the idea of a 3-dimensional world down our own throats. Now we have these no-mass things being further reduced by "vibrational cancellations". All this to try to hold onto the 3-dimensional universe concept. Why don't we simply decide that the Universe CANNOT be 3 dimensional? Here is what Einstein said about the stuff.

We may assume the existence of an Aether; only we must give up ascribing a definite state of motion to it, To deny Aether is ultimately to assume that empty space has no physical qualities whatever. The fundamental facts of mechanics do not harmonize with this view... According to the general theory of relativity space without Aether is unthinkable; for in such space there would not only be no propagation of light, but also no possibility of existence for standards of space and time, nor therefore any space-time intervals in the physical sense."

I know all of this is confusing, but for us, the only thing we need to take away from this whole concept is that everything vibrates. If vibrations are in opposition, they cancel matter. This happens in the normal world, the microscopic and macroscopic world.

## Vibrations

Vibrations of nothing are not three dimensional. They are one dimensional as they have no mass. Just clear your head and open up your mind so that we can re-identify the universe. Not one filled with anomaly after anomaly, but one in concert with reality.

The simplest form of reason associated with our universe is a VIBRATIONAL organism, not filled with objects and volumes, but filled with huge numbers of vibrating nothings.

To keep from having to say nothings we can call the things fields, but before we can get to be a field that can affect a surrounding there is a lot of work to do.

Don't be scared? You are up for it, and I will try to make it interesting as it is very important for understanding all of the following:

*Heaven, Life, Consciousness, Light,*

*Miracles, Death, Spirit, Soul,*

*And what a rock is*

If you want to know about these, we need to change our concept of the universe and reality. Instead of having one answer, here we are going to have to investigate three different major concepts.

- ***Vibrational Matter*** *in a 10-dimensional world*
- ***Transverse Time*** *allowing backward motion of time*

- *And finally, something I like to call **transposition of the universe**—one that is frequency based rather than the more easily envisioned time based one.*

I know these things sound obscure, but I guarantee that your eyes will be opened to our universe, religion, heaven and all the rest.

## Not New

As I mentioned already, the whole 3 dimensional universe was falling apart. I purposely left out the nasty part called quantum mechanics, because it required space time to not exist and without space, there could not be volume. Without volume, there could not be height, length, width. Let's just forget about a universe based on volume and mass and look at one more, very important anomalous characteristic that is required for our universe to exist. This was investigated by Albert Einstein and others. It is called consciousness. It seems that our consciousness changes reality. Einstein called this control relativity. Let's see how that is going to help us draw a picture of the universe.

Before we can understand how to modify our life resonance, we must understand what it is. Unfortunately, the next 30 pages will be somewhat physics based, but I will get our of it as quickly as I can.

# Relativity

Let's start by saying that Relativity is a confusing part of reality. I'm not saying the other elements are not confusing; what I mean is that this goes way beyond Length, Height, and Width. We know about it instinctively, but typically don't put it into worlds.

Einstein told us and we have later confirmed that as we go close to the speed of light, our size reduces and time almost stops for us. Let's say two people are together and the same age. One goes traveling at close to the speed of light for 20 years. Neither shaves until the traveler returns. When he returns, the fast guy has not aged a day and the other guy is 20 years older and has a 20-year beard, as shown in the following diagram. It didn't matter where the traveler went and how he came back so long as he did it close to the speed of light. He could even stop every once in a while to see stuff. His reality changed while he was spinning or traveling or whatever he did if he did it really, really, fast. For the guy going fast, time almost stops, just like the suspended animation "sleep" method we see on science fiction shows.

To explain it, we simply used what Einstein referred to as "relativity" which forces the speed of light to be constant to all observers. If you are going the speed of light, you must experience light "generated by you" going the speed of light, so you must not experience the time during your travels AT ALL.

## Relativity Proven

Of course they have proved this anomaly with a number of experiments, but again, this only allows the forward dilation of time and there is substantial evidence that people can go forward in times as well and still become part of "normal" reality once the journey is complete. If you went on your "almost the speed of light" journey for a thousand years, everyone you knew would have gone. That would be a bummer.

Notice I showed that the guy going close to the speed of light does not have to travel in a straight line. Let's see where that takes us.

Instead of going somewhere, the fast guy simply turns in a circle, the same thing happens, in the blink of the fast guy's eye, the slower guy turns old and feeble.

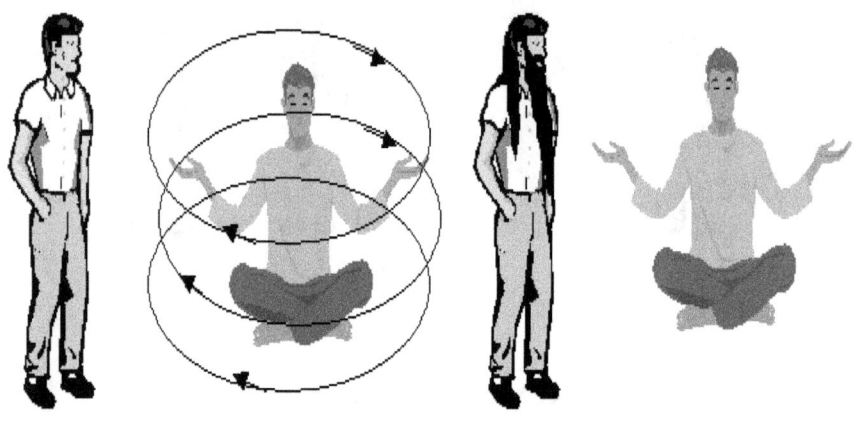

Let's look farther. Let's say the spinning guy turns on a flashlight aiming outward. What would the slow guy see? Each time the flashlight came around in his direction he would be blasted in the face with light. Because of the spinner's speed, that would be all the time. To the slow man, he would, sort of, see his friend turn into a light as shown next.

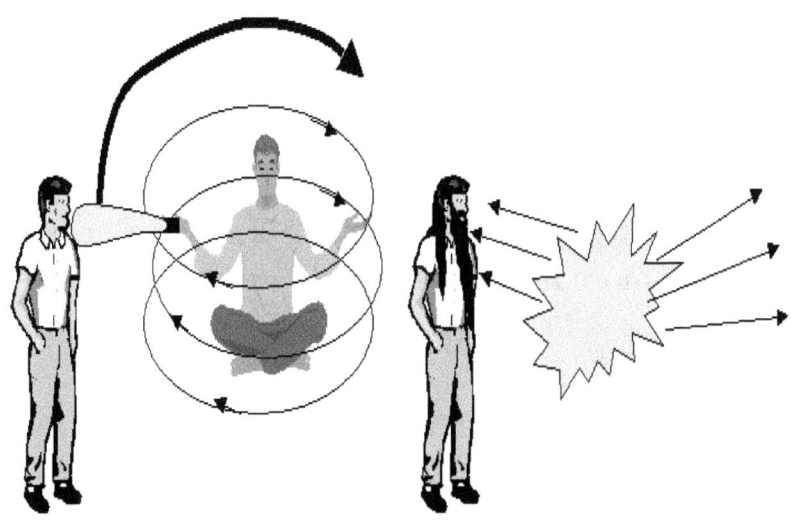

**Tiny**

I forgot one more thing in the example. As the fast man spins faster he gets smaller and heavier. Close to the speed of light, the spinning man is very small and very heavy. Mass tends towards infinity as the object approaches the speed of light.

Now assume that the spinning man was not just moving around in a circle but was spinning in all directions and you see that he becomes a speck of light. The next time you see light just think of it as **a spinning man with a flashlight** and time travel may become a little better understood. OK! That was a joke, but that is what modern science is wrestling with as the 3-dimensional universe was shattered.

**Vibrate**

Do one more thing for me. Think about what would happen if the man didn't spin around in a circle, but instead, he stayed still and his body vibrated close to the speed of light. He would again stop aging and time would almost stop for him. He would decrease in size and increase in mass---all while he was standing in front of the other man. The vibrator had actually traveled into the future or the standing still guy. In a way he messed up time/space. Just think about that for a minute and let me bring up one more nuance about this relativity thing.

### Tree Falling

This is an old question and I mention it in many of the vibrational matter books! If a tree falls in the woods, does it make a noise? *"**Einstein** said, if no consciousness is involved, there IS NO TREE to make a noise. We must have life and consciousness, or the entire universe could not exist."*

### Now What Do We Do?

This consciousness thing has no mass and is not electro-magnetic. It is something different. To make it more confusing, ancient Jewish texts talk about this life/consciousness as something they called "the light". I'm not going to present them in this book so that you won't get confused, but I do want you to know that is someone wants to walk across water like Jesus, Elijah, Peter and the others from Biblical times, one can do it. If someone has something called faith he can move mountains as stated by "God incarnate" or Jesus. This is not outside physics. It is simply outside the height, length, width physics that has been shattered over and over again.

Life itself is a component part of our universe. It is one of the dimensions REQUIRED for our universe. Hopefully, you are starting to see that our universe is defined by something that looks like matter, something that looks like light and something that envelops something we call consciousness and ALL of these things can be characterized as vibrations of nothing.

## God Incarnate Turned Into "A MAN"

I know this stuff sounds stupid when you first read about it, but please try to continue and you will begin to open you eyes to a new world. One where religion and science can coexist and Jesus could come from a universe called Heaven and become a man. In our universe Jesus could, as a man, could walk on water and make water into wine. While he was also God incarnate, he had to be 100% man or Christianity would be a false doctrine. Because of these new truths, one can believe in religious teachings without conflict.

With that, let me introduce you to what string theorists call dimensional strings. While many don't see the key that connects the 10-dimensions together, I think this will begin to make sense to you as dimensions are not straight lines called length, Height, and width. They wiggle.

# 10-Dimensions

I know I have been skirting around the question about how our consciousness affects our reality, but I'll be getting into some of it right now and if you get a chance, please read my book on "Our 10 dimensional Universe". I am sorry for this section, but to understand how one might change his vibrational resonance, one must at least accept the various dimensions that make up a Vibrational Universe like the one we are experiencing. It will help clear up some of the characteristics presented and maybe even allow you to see that religion and science are talking about the same thing concerning how consciousness affects reality and just about everything else. Below are some of the things I think you may have sensed.

- You have this burning idea that <u>gravity and magnetism have a similarity</u>, but both appear to be completely different. In fact, no one has ever truly defined what gravity really was in the first place. Magnetism is sort of the gravity of the electro-magnetic things of the universe.

- Gravity has nothing to do with length width, height or time, but you're sure that gravity does exist and it is REQUIRED for our universe to exist. A mass without gravity would certainly be odd.

- Sometimes Light is a wave and sometimes it is a particle. That is not the answer. A different answer might be easier to understand. One might believe that energy and mass are the same thing "at times".

- The scientist guys came up with something they call Nuclear force, but you can't understand it with 3-dimensional space. Atoms should not be holding themselves together. I know scientists came up with these GLUON things, but Nuclear energy is simply the potential to make MASS, just like Electricity or voltage is defined as a potential to make electro-magnetic things do something.

- Magnetism and Electricity both work for us all the time, but their existences are not covered in the normal universal definition. Certainly, no one would try to indicate that either of those items did not exist.

- We also must consider life itself. The universe without life might not exist. For us, we can certainly say it would not exist because we would not exist. Life is not DNA so don't go thinking it's covered by a "volume" of particles.

**How about our consciousness?** Does our consciousness exist? If it does exist the question might be "If I had no consciousness, would the universe cease to exist?" Some will say that people being here or not being here have no affect of the world at all. This type of thinking is not correct as more and more observations that only were understood by Anthropics are found every year. This "consciousness thing" is not governed by the 4 normal dimensions so we need to investigate how it is constructed in this universe if we have any possibility understanding life.

**What about LIFE itself?** The question would be, "If there were no life, would the universe exist?" The cop-out answer would be, "Just like consciousness, a universe without life would not be in existence."

Luckily, in a vibrational world, we are not governed by the constraints of 3 dimensional volume that was shattered by the discovery of Quantum Mechanics which builds around something called a frequency domain. Instead of a particle world, a frequency domain world allows and attracts multitudes of volumetric entities as working units inside the frequency domain. Many of the volumetric elements have a frequency that is similar and they want to stay their or be resonant. This general resonant vibration is perceived as reality. We can leave time to its own devices [for now] and define a truer universe. As I mentioned, most scientists indicate that all but 4-dimensions are <u>compactified</u>. That is they exist, but they are "discarded". Why would there be dimensions that don't exist, but do exist????. My feeling on this is that these string theorists that defined this <u>compactification of dimensions</u> are nuts!

Let's try to find the dimensions. We'll see they are compactified at all. It's not hard, but you must think of the key dimensions as vibrating strings so that you can picture them.

In a Time-space perceived universe, it is quite easy to <u>ignore</u> key dimensions that are required to continue an existence [life and electricity, etc.] In a frequency based, "truer" existence, some of the dimensions are not as easy to ignore. In the first place, items defined as frequency based now become critical in our universe rather than simply being characterizations of our universe. Let me reintroduce the, now important dimensional elements as three groups Structural, Operational, and Ethereal. They don't have to be called that, but they seem to fall into these categories fairly well.

Once we recognize how the universe is built and stabilized, we can begin to change it to make our life better.

# Structural Group

Instead of 3 dimensions being controlled by a 4$^{th}$ one [time], our universe and our life is governed by 3 sets of 3-dimensional groups. The Structural group could be considered the length, width, height dimensions, but those require time and distance which Quantum mechanics and relativity destroyed. Instead, a particle has been defined as a "Standing Wave" [intersection of 2 or more vibrational waves] which vibrates at different frequencies larger particles vibrate faster, smaller particles vibrate more slowly. Its "resonant vibration" determines what the particle or object is and it is governed by 3 dimensions.

### Fermionic Dimension

It is this dimension that allows the creation of matter when matter cannot be created. I know you thought Stephen Hawkins was crazy when he indicated that matter could be created at something he called the event horizon. Well, this dimension sort of allows that action by transferring matter in accordance with the law of super-symmetry.

Fermions are the quasi-particles that vibrationally associate with others to form bosonic particles which build to Baryon particles which become Atomic particles, which become molecular particles which eventually make everything that is Matter. While fermions actually affect our universe, the base of this dimension is something called Aether [according to Einstein].

Structurally, most of the universe is made up of Aether. **Aether can be viewed as the potential for having matter**. Once Aether begins vibrating it can become fermionic and if it goes fast enough it can become an atom. The larger the atom the faster it vibrates until it becomes something we call a black hole that as almost completely something we call gravity so let's look at a gravity dimension.

## Gravitational Dimension

Gravity the dimensional quality of matter that is perpendicularly associated with the fermionic substance of matter. We know it is around and even the things we define as fermions usually have some type of gravity component, so don't go thinking gravity has something to do with mass. Gravity can be defined as the operational construct of matter. It holds matter in existence and that brings us to the force of matter or what I call the Nucleatic dimension.

## Nucleatic Dimension

It actually is the combination of fermionic dimension and the gravitational dimension. One way to look at this dimension is to consider Nuclear attraction as the actually vibrating interface to matter in an adjacent universe. The graphic following shows this characteristic of our universe in 2 ways.

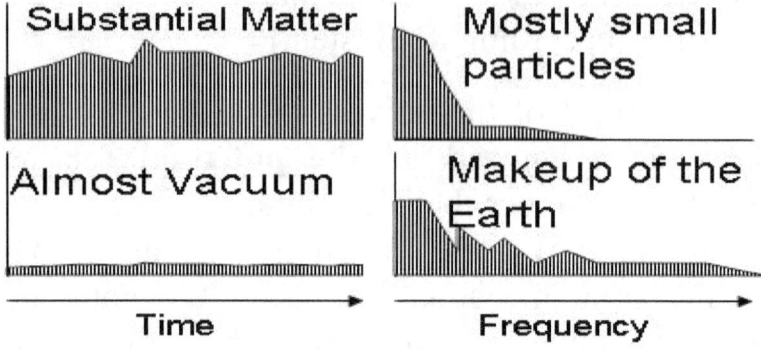

The first graph shows how structures change with time. Beneath it is what we might expect from a vacuum-almost no energy at any time. It changes a little, but in general to us a desk always is a desk and a vacuum is always a vacuum. The second one shows the vibration spectrum one might record as an average in the universe. As most particles in the Universe are small, they vibrate at very low frequencies as shown. Notice that on Earth, some of the heavy metals can be found which gives Earth more significance [vibrationally] as more energy is provided by higher frequency particles just like higher energy is provided from higher frequency electro-magnetic waves. Don't worry about the graphs, they are simply here to show that we can better measure our universe and ourselves by vibration to get a feel for how we affect the universe.

# Operational Group

If that was all there was, the law of entropy would soon convert all particles into the lowest level of vibration which is something less than a particle that Einstein called the Aether. The Operational group could be considered the force dimensions. Later we will discuss where the forces come from, but right now understand that the electromagnetic, kinetic, potential, rotational and even life stresses all are made with these dimensions. Like the other structural group, an operational group vibrates at different frequencies more force vibrates faster, less force vibrates more slowly. Its "resonant vibration" determines what reaction or force will be exerted and it is governed by 3-dimensions.

### Electrical Dimension

This dimension is characterized as a phase shifted, vibrating magnetic field. Another way of looking at it is something we call voltage or **POTENTIAL to manifest any operational force**. Pure electricity is the lowest frequency force as it is simply a potential to create a force so it does not vibrate by itself.

## Magnetic Dimension

Magnetism is a characterization of vibrating electric fields. Actually, the magnetic fields are always perpendicular to electrical fields. One could say true magnetism is the true operational construct of the electro-magnetic spectrum. Speaking of electro-magnetic field resonance or effect, it can be viewed or measured in the $3^{rd}$ dimension of this group.

## Photonic Dimension

What we call photons are actually vibrating waves of electro-magnetic fields. They are the electromagnetic interface to another universe. This allows the make up of light to change drastically as frequency is modified by swapping energy with an adjacent universe. One can say photons can be characterized as Electro-Magnetic Force or the effect of electro-magnetism.

As with Particles, we could plot Electro-magnetic [EM] fields [photons, radio etc.]. The graphs following show the difference one might expect in plotting energy to time and energy to frequency. The Frequency graph is more important and is shows how high frequency, high energy EM fields might be able to alter our universe. The same will be true in the ethereal group of dimensions.

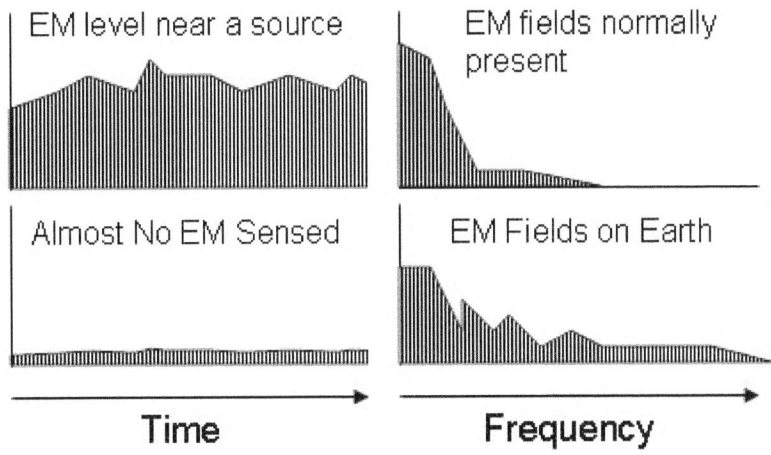

Like the other set, EM measurements don't change with time unless acted on. But when we view the vibrational world, we can see something interesting. An average EM resonance curve might be shown as the 2nd graph while the last graph shows that the "normal" EM measurements would be perturbed by increasing the resonance of particles [making them larger] and by changing the resonance of something we may call the collective consciousness. We will be studying this one.

The limit of electro-magnetic force is complete magnetism just like the limit of particulate force is that black hole that is totally gravity. As electro-magnetic forces increase in vibration they become more and more dangerous to us and the universe and we can measure the resonance level of this 3-dimension group and use them in radio-waves, light, X-rays, electronics, and Gamma rays. Cosmic rays are the fastest measured resonances, but who knows. Perturbations' in the EM resonance changes reality.

# Ethereal Group

That brings us to the ethereal group of dimensions or [LIFE]. The dimensions of life have a similarity to the other 2 dimensional groups.

### Self Dimension

"Self" itself is made up of vibrating chemical responses set to some generally unknown master clock. It isn't the proteins of DNA that are life it is this vibration thing.] Some call this dimension self awareness, or the EGO, but the thing to understand here is that there is a continuous struggle in ones life between this key dimensional quality and the other 2 ethereal dimensions, what I call here consciousness, and Spirit. **While no one has defined life well we can say it is the POTENTIAL for forming conscious-Spirit force.** What I mean by that is that life by itself does nothing to affect the universe just like Nucleatic and Electrical Dimensions. When acted upon by the Spirit, a consciousness force can be affected. Things like bacteria and insects have life, but without consciousness, they cannot affectively change reality in any way. They are just there just like a rock.

## Spirit Dimension

While consciousness is very difficult to grasp initially, the dual of Photonic and Fermionic dimensions is this "spirit" thing. One could say this is the "Key" to transferring life between universes. This is going to sound weird [I know all of this is weird, I was just seeing if you were reading the information], but think of this as a cause and affect because "spirit" reacts to "life" like "magnetism" reacts to vibrating "electricity", or like gravity reacts to matter. By affecting the vibrational level, "life" is changed and visa-versa. If a life vibration is halted, "spirit" must find another life to be able to sustain itself in this universe. The whole concept of angels appearing on earth as they transferred themselves for heaven is not a myth, nor does it violate some physical law. This strange dimension allows the creation of "Conscious-Life" just like the Gravity dimension allows the creation of matter. The Spirit can be defined as the operational construct of Life-consciousness. It establishes consciousness in life.

## Consciousness Dimension

Consciousness is not exactly life but its vibrational essence is perpendicular to life. Some call this dimension the Soul of a person, or the ID. Some define this as the little voice in the back of your head or premonition or self actualization characterization. What we call Consciousness actually vibrating quasi-particles of Life-spirit fields. This allows the make up of our life-spirit to change drastically as frequency is modified by becoming more or less Carnal centric. It's sort of like the more spiritual you become, the less carnal or "Life-only" you become. One can say Consciousness can be characterized as Life-spirit Force or the effect of life and spirit working in concert.

I know I went spooky on you for a moment, but as you read this book please understand I am not trying to devise a result. I'm simply trying to look at the stresses on the universe that make what we call reality in a little different way than you have sensed it in the past. I will be doing it vibrationally rather than as referenced to time. I will be transposing the characteristic of the universe so that you can see it better and interact with it better. If we had a reasonable way to measure life frequencies and we tried to gain knowledge in a graphic way, the following graph may show what life might look like. In this graph, we can't tell the frequency, but we can se if a life form is vibrating. The more vibrations we pick up, the more life there is in an area.

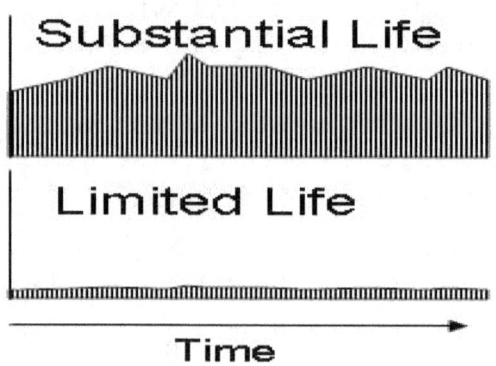

### Elijah Walked on Water

While that has some use, there is no way to determine what is going on with that life. Just like in electro-magnetics, a time map is substituted for a frequency map to show us a better picture of life. Notice that basic life is mostly centered on low frequencies associated with survival, sex, and self, while human life has a small amount of frequencies that extend well above the basic animal characteristics. <u>Getting to those frequencies effectively is how you change the universe.</u> It is how someone can raise a car off a small child. It is how Elijah walked across water.

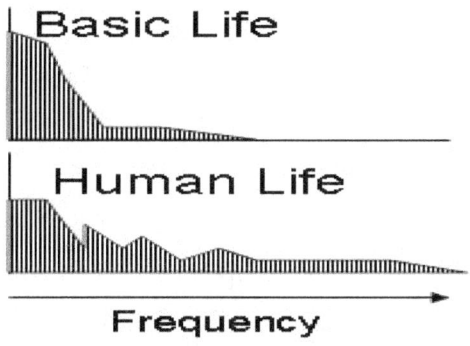

The way to truly affect the universe is to change how fast your consciousness vibrates. Faster vibrations allow for more energy transfer and control in our universe. People who are into deep meditations seem to affect this vibration. We are told that a high pitch tone can be heard as once meditates away form the base sex, self, survival level of existence. We can easily measure change in the operational group as electro-magnetic waves can be intercepted, tested, measured, manipulated, and modified by changing their resonance characteristic. It is not so easy to measure resonance of the Ethereal Dynamo. We must concentrate on effect rather than normal measurement. As we "resonate" at a higher frequency, one can control his "reality" more. Given enough control, one could move a mountain just like Jesus told his followers they could do.

While it seems difficult, one can learn how to affect his life vibration levels and extend his control over his life, happiness, and power. While it will be easy to tell you how, it will be difficult for you to accomplish the task. That being said, isn't life, happiness, and power important enough to do it? By the way, when I say power, it does not mean power over other people. Trying to do that will certainly limit your growth. What I mean is power over you environment---your universe. The first step is simply ignoring your selfish "need", sexual desire, and survival fear.

## Grunting

I don't have any way of knowing if you are reading this and grunting or if I have extended your awareness a little, but the review I just presented is needed for us to go along. As you can tell, there are three distinct dimensional globs [or dynamos] that I have described. If any one of them is removed, ALL of the Universe stops. That is how you know they are dimensions. That is nine of the dimensions and the 10$^{th}$ "may be" what we call time.

*The entire universe can be express by 3 distinct things- structures, forces, and life.*

Instead of the normal 3 dimensions with time as the activator, our world is really split into 3 sets of thee dimensions. Each set of three works together in a similar manner as you had originally believed the 3 dimensions reacted, but separated along lines associated with vibration rather than spatial separation. As it turns out, the separation is an effect rather than a dimension. Let me give you an example. Length and width and height are all exactly the same dimensionally attributed components that are mutually perpendicular. While perpendicularity is a sign of dimensional separation, the reactions that cause perpendicularity must be reexamined. In a truer world matter is defined by spatial separation, gravity, and nuclear bonding. Length, width, and height cannot define all of these things, so a different definition must be considered. In a 10 dimensional universe all dimensions must somehow be perpendicular to one another or they cannot be defined and other dimensions would affect them.

# Mutual Perpendicularity

In this universe, everything that interacts but holds a definition unto its own must be mutually perpendicular to the other defining elements. What I mean by this is that there are key elements that make up the universe. While they may have some similar characteristic, they also can act separately in the universe. It's that way with the length, width, and height elements and in the vibrational defined universe, the same thing holds true.

### Mutually Perpendicular God

If you remember from Sunday School it is that way with the Creator God who defined himself as a dynamo of thee entities "Father, Son, Holy Spirit". He was describing the 3-dimensional dynamos that make God and man. God's dynamos are simply overarching and our dynamos are subjugational.

We would not be able to understand God in any way except as three emanations of three dynamos of existence personified in the ethereal dynamo as self, spirit, soul. Of these, we can sort of imagine the self and spirit portions, but the soul of God is well beyond anything we can imagine. There is No mystery in how God can and must be three entities and [oh boy! I'm going to get in trouble for this one] people must be defined the same way. We are presented in the universe as a 3-"person" entity- the Operational entity, the Structural entity, and the Ethereal entity. We don't get confused in our being because each characterization is mutually perpendicular to the others. Each of the 3 is sort of synchronized by the others and, generally inseparable. The physical/structural entity can be easily understood and the operational entity that allows us to move around and do things can be easily observed, but the ethereal self is going to take some time.

The advantage we have with mutually perpendicular dimensional dynamos is that vibrational elements are more dynamic than the original, extremely limited definitions of our universe. The diagram following characterizes the 3 dynamos and their individual perpendicular dimensions joined by the dimension of time. Of course, this is not the way it actually looks. All of the dimensional strings would look sort of like vibrating circles, but it does describe the connective nature and the "free" dimensions called life-consciousness, Electro-magnetism, and Fermio-Gravitation. If "dimension" makes you uncomfortable, you could also say Spirit Force, Photonic Force, and Nuclear Force.

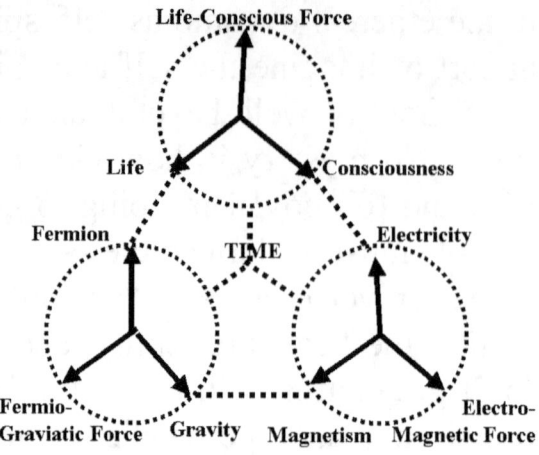

The "spirit-force" is the one that is of most interest here, but let me get back to definitions.

## Why Three Dynamos?

As I mentioned, modern string theorists tell us that there must be at least 10-dimensions that define any universe. Mathematically it makes sense, but even more importantly; the characteristics of our universe cannot be defined with less.

By grouping the dynamos into 3 groups, each of the dynamos can also be mutually exclusive with respect to the other 2 dynamos as well so that the 3 dynamos can reacts exclusively to themselves and all dimensional elements can use the same time activator as the 10 dimension. With that, I know you are confused so let's go on to time and get straightened out so you can increase you resonance and have a better life.

# Three Secrets of Time

As shown in the last diagram, time, sort of, connects all three dynamos together. Guess what! The dynamos or their resonances are mutually perpendicular to the existence of time. We can say that time has three mutually perpendicular "dimensions", but in this case, the things that make up time or the things that time establishes are the three dynamos or dimensions.

One could say that time is not exactly a dimension, but, it can be characterized as one and I think everyone can recognize the dilemma our universe would be in without this critical dimension. Because **God senses time differently than we do** [we are told he sees the beginning and end of time simultaneously], we cannot easily compare God to man when time is the controlling element. Let's just stick with mankind sensations right now.

What we will find out later are 3 very important secrets that we have not been taught, but we must try to understand. Time MUST BE constant rather than always progressing. What I mean by that is if time is always going forward for us in this universe, **there MUST BE a timeline that is opposite to our perceived time** or time would eventually extend beyond the limits of our universe and everything would start slowing down or the characterizations would become less dynamic until no time could be generated. Don't worry about that right now, it will take some explaining. This factor is the main reason that a transposed vibrationally defined universe is so much better than a time defined one we are more comfortable with.

The second thing to generally be aware of concerning time is that each entity perceives his own time. That perception is more than simply believing time slows or speeds up, time for that entity ACTUALLY changes during the course of a day, week, or year. The easiest example is the one that Einstein presented with the person moving fairly close to the speed of light. He not only perceives time is going super fast, to everyone else, time had gone super fast to the almost light speed guy and he doesn't age. This rule has been proven so don't go thinking Einstein was a nut. You do not have to travel fast for the phenomenon. If you play an intense game of tennis, time seems to and actually does go faster for the player. If one watches a clock, the opposite "actually" does occur to the observer.

There is also a third element of mystery and that is time seems to be controlled by a collective consciousness. We will get into that in later books as well so don't even think about them right now. Simply begin to see how the various dimensions are interconnected and still mutually perpendicular to allow "free" modification. The modifications result in resonances which DEFINE how we perceive the universe and how we can control our universe. Because a collection of conscious observers are joined by something we could call "life resonance leveling", we all generally see, feel, understand, and react with our universe in the same way. Before we get into leveling, let's look at resonance itself.

# Space Resonance

Dr. Milo Wolff begins this discussion. [my comments are in bold]

*"Resonance is composed of a spherical IN-wave which converges to the center [of the universe and comes from a different universe as a component of the operational dimension dynamo] and an OUT-wave which diverges from the center [of the universe and makes up what I call the structural dimensional dynamo]. Their separate amplitudes are [close to] infinite at the centers. [Like all other resonance factors in the universe, how close they are to being infinite can be considered the "quality of resonance".] When combined, the two waves form a standing-wave which has a finite amplitude at the center. The standing wave [appears] to be the structure of the electron. The inward and outward waves [sort-of] provide communication with other matter of the universe. Spin of the electron is a result of the reversal of the IN wave at the center to become the OUT wave."*

While there is more to this, the understanding that universes are held in place by resonance will help you see how significant this is. Milo was speaking of operational and structural dynamos, and I will show you that the Ethereal dynamo acts EXACTLY like the other two. As one bit of information that is of paramount importance, the more we communicate with an adjacent universe the higher in frequency our resonance becomes and its quality rises.

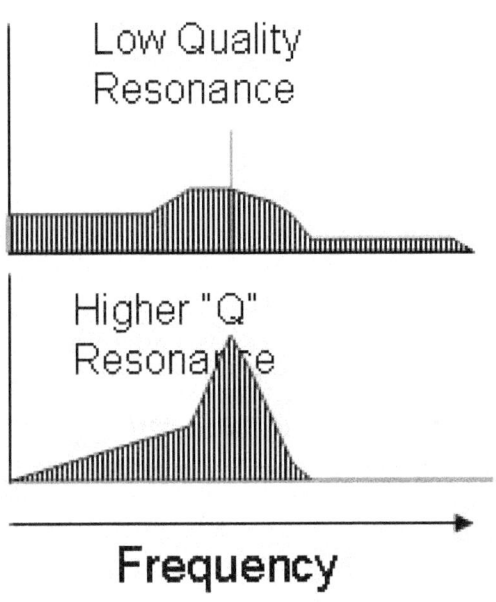

# Quality of Resonance

Let me explain this "quality of resonance" a little. One way to do it is with something we can measure a little easier. In electro-magnetics, quality of resonance describes the difference between the effect of a circuit outside its resonance frequency and that which can be described when it is in resonance. Let me explain it this way. If a crystal is excited with a vibration that is half of the frequency it likes, it may vibrate a little and nothing more, but if it is hit with the vibration it likes, it begins to self oscillate substantially unless it is weighed down by other devices around it that don't like the resonant frequency. The quality of a resonance is generally controlled by the purity of the environment or how well the device resonating can ignore the other objects around it..

It is this reaction that describes "quality of resonance" for ANY dimensional Dynamo. It is how well the device, or field, or consciousness responds to critical vibrations and ignores others.

In the electro-magnetic world, this quality depends on many things including what the crystal is attached to, how well the crystal is cut and how homogeneous the crystal is. In the electron or particle world, the same things can be surmised. Purity of the particle and the things that surround the particle affect how close to infinity the standing wave appears. It's funny that in the operation dynamo the word purity is so very important in its control over the universe. The Purity of Conscious acts the same way. Purity in the ethereal dynamo is measured differently in that what we consider Carnal—That which we interpret to be the lowest level of existence feels the most natural while spiritual endeavors seem the farthest from CARNAL living and that appears to be exactly the way to view pureness or the affect of "Quality" here. The graphic below shows what the vibrational spectrum of someone gaining purity of consciousness, sometimes called Self Actualization. Basic animalistic low frequency elements are replaced by focused higher frequencies that can affect the universe. By removing the low frequencies around you, your consciousness can resonate at a higher frequency associates with controlling your destiny and your reality. I know that sounds just like words, but just think about it and don't ignore what I'm telling you.

When Paul [from the Bible] was asked if people should marry, his answer sounded sexist and barbaric, but he was talking about the same thing. He said "Would that you could live like me (celibate), but if you burn with lust get married" (so lust will not bring you conscious vibrational level down and keep you from enjoying control over your environment) . OK! I added the last part, but that is what he was trying to tell people. An example is shown below. Even life resonating at a higher frequency because of some deep devotion to others, the more basic sex, self, survival keeps the resonance from making a huge affect on the environment.

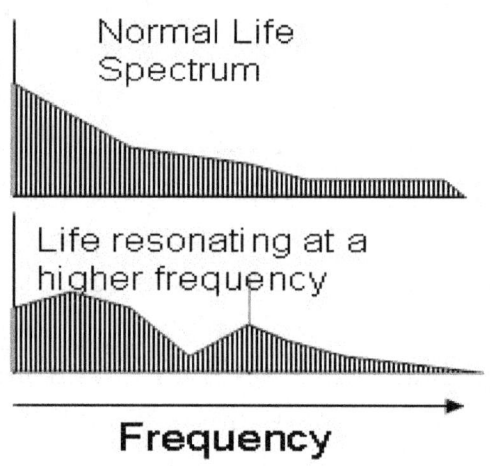

I know the graphics are lame, but please just try to get a sense that everything can be defined more vibrationally than with respect to matter. As you begin to expand your awareness of life's resonances, you will actually be able to feel the vibrational levels of your body getting faster. That is a sign that you are getting closer to being able to affect your place in the world.

Here is one more thing to notice. The people who really seemed to be in control of their reality are almost always loners. The 3 "Ss" can't attack a easily when you are alone.

I'm not saying be a hermit and enjoy control over your reality. You could just as easily go mad. If you want control, you have to sort of give yourself away and understand everything and everyone's deepest problems. Once you understand, you must try to make those issues yours. It's kind of like empathy with a huge EMP.

# Life Resonance

I know all that stuff was difficult to get through, but I needed to open your awareness to a truer understanding of our REALITY than get from people who want to limit you. I want to expand your awareness and capability in living a satisfying life and even a better death. To be sure there is a massive secret here so don't pass it on. Consciousness/life/ and death all act the same way. If we wish to affect the universe more and have a higher "quality" of resonance, we must become pure and surround ourselves with things that allow this pureness. OK! I can't exactly define what pureness is, but all this prayer and meditation stuff is probably more important to our quality of resonance [and our capability of affecting the universe] than one would initially believe. Every time one thinks about himself, pureness is lost. Anytime one tries to affect something by "reaction" rather than true understanding, pureness is reduced. Any time a person tries to work on his own, pureness is weakened. To achieve pureness, one must become what Thomas Maslow called "Self Actualized" and that is just the first step.

# Self Actualization

As a first step, let's investigate this term "Self-actualization". It has been around for many, many years, but it was brought into prominence in Abraham Maslow's "hierarchy of needs theory". It was the final level of psychological development that can ONLY be achieved when all basic needs are fulfilled and personal "actualization" takes place. After investigating a large number of successful individuals that seemed to all have similar characteristics, Maslow determined the following about what he called Self Actualization:

***Efficient perceptions of reality*-** *Because they are not focused on self, self-actualizers are able to judge situations correctly and honestly. They are very sensitive to the fake and dishonest.*

***Comfortable acceptance of self, others, nature*-** *Self-actualizers accept their own human nature with all its flaws. It becomes unimportant to them they sort of go outside themselves. The shortcomings of others and the contradictions of the human condition are accepted with* humor and tolerance. Like those of himself, the carnal character means less and less.

**Spontaneity-** *Maslow's subjects extended their creativity into everyday activities. Because they were no longer internalized, they tended to be unusually alive, engaged, and spontaneous. It is this lack of importance that will allow the quality of resonance to actually increase.*

**Task centering** *-Many of Maslow's subjects had a mission to fulfill in life or some task or problem outside of themselves to pursue. One way to limit carnal life is to focus outside one's self with a massive mission. Once the goals were set, real change in that mission was noted.*

**Autonomy Self actualization** *causes resourcefulness and independence because self has little meaning. The goal becomes the person.*

**Continued freshness of appreciation** *-By moving outside one's self, one gains a profound appreciation of a sunset or a flower. The "innocence of vision", marks the true nature of self actualization.*

**Fellowship with humanity** *-Empathy is the most pronounced effect of self actualization. Being able to feel a deep identification with others is of utmost importance to control the universe.*

**Profound interpersonal relationships** *-This is not the NORMAL lust driven love, or even the normal love we consider to be even closer. This is love intense enough to promote losing one's life to protect another type love.*

***Comfort with solitude-*** *While self is very low on the totem pole, once one is self actualized, there is no issue with aloneness. It simply is fine.*

***Non-hostile sense of humor-*** *This refers to the wonderful capacity to laugh at oneself.*

***Peak experiences-*** *What was reported mostly was that self actualization could never be considered a continuous thing. Instead his subjects reported temporary moments of self-actualization, marked by feelings of ecstasy, harmony, deep meaning, and, most importantly, a feeling at one with the universe and filled with light.*

Maslow determined that the Killers of self actualization were the primal characterizations of man. Form that he derived the "five needs of man"

*Physiological need*

*Need of safety*

*Need to belong,*

*Need of self esteem,*

*and finally self-actualization*

My list is somewhat different, but it has a level of similarity with the lowest frequency elements listed first.

*Need for Survival*

*Need for Sex*

*Need of self awareness*

*Need of companionship*

*Need of helping others [Self Actualization]*

*Selfless love [to the exclusion of self]*

*Understanding of the universe*

*Understanding outside the universe*

If you have any chance at affecting the universe, you had better get self actualized or temporarily blast through to this all important step. I'm just going to leave this discussion right now and get back to particles so we can better define this resonance thing.

## Resonance and Matter

The truth is that "resonance" holds matter together and it holds time together. If enough electrons are in an area that are sensing similar in-waves, they align together just like a crystal. One could say that atoms are resonant plugs that are held together by like-vibrations. Scientists found these things called gluons which seem to act in opposition to other particles and quasi-particles. Gluons hold quarks together and 3 quarks and an unknown number of gluons are called an electron. Gluons are quasi-particles [fermions] that have a negative gravity. That is; the farther the quarks move away from the gluons the STRONGER the gluon attraction becomes. It is sort of like the quarks are inside and invisible piece of matter that has a gravity. The closer they get to the surface of this invisible piece of matter, the more the gravity affects the quarks. The center of this gluon, matter would be the resonance point of what we call the electron. Its resonance is defined by the vibrational characteristics of its component parts.

Gluons are not odd, they are simply invisible. One can say that they are this in-wave out-wave collision.

## Who Cares About Resonance?

Why have I even brought up resonance? If a vibration node gets larger or smaller shouldn't matter to us. Right?

*Well----we need to care for a number of reasons. Here are a few.*

- The higher the resonance of electro-magnetics, the closer it comes to being light which is the most stable electro-magnetic form and the most useful form of electro-magnetics to the universe.

- The higher the level of resonance of a particle or quasi-particle, the more stable the matter it establishes becomes and matter is the most useful form of a quasi-particle to the universe.

- The higher a person's resonance is, the more he or she can affect the universe and their own characteristic universe becomes more in sync with everything else. The closer one can get to God.

- In order to hold onto a resonant state, the quality of the resonance must be increased. <u>Self actualization</u> or chanting, or other ways people use to think of existence outside our basic needs increases the quality of resonance. And vibrating faster makes our resonance more controlling. All these spikes in vibration levels are sort of moderated by those around us.

## Resonance and Higher Frequency Vibration

When I talk about resonance, what I'm really saying is that we need to increase the resonance of Structural, Operational and Ethereal elements of life as we also try to increase the vibrational frequency. We can't generally establish larger particle, increase Electro-magnetic vibrations to very high levels, but there might be something we can do with our consciousness. If we increase our vibrational frequency, the entire universal resonance level at these important frequencies goes up and it subtracts from the lower frequency, animalistic elements of our existence with something that can be called Life Resonance leveling.

# Life Resonance Leveling

OK! I brought up a new term for those who haven't read the "10 dimensional Universe" so I had better define it a little. Like EVERYTHING, all dynamos seek it lowest level of stress. Some call this entropy in a particle world, but the same thing happens in the electromagnetic world by reducing everything to an electric potential. Well, the same thing happens to our lives as we continuously will be debased to our lowest need levels of self awareness, survival, and Sex. There are outside encounters and internal aberrations that modify the various resonances. As these "resonances" change, the concept of time changes. These resonance qualities cannot be mapped in a time based world easily, but when we transpose the measurement scale to frequency when testing electro-magnetics, the characteristics can be easily sensed and the outcomes of the variation could be foretold.----

In the particle world, when testing for resonances of structures, one can easily sense it by looking for densities of atomic structures. Denser generally means higher frequencies and having more mono-particulate areas tends to show Quality of resonance. What I mean by that is the heavier atomic bunched together can control the universe more than a clump of Hydrogen or helium. An example might be nuclear fission when too much uranium is placed in close proximity to each other. UNFORTUNATELY, while frequency of electro-magnetics and even particles can be measured, vibrational characteristics in the ethereal dynamo have no easy measuring tool. Some examples of where ethereal resonance quality is high might be the following:

- *Man raises an auto off a small child and is not hurt. Others jump in and help.*
- *Peter [from the Bible] walking across the water when Jesus was walking on water.*
- *Man thinks incessantly about becoming rich and all of a sudden he becomes wealthy as do his friends*
- *One man dies to save another. Others do similar acts nearby.*
- *Other acts of heroism, especially group efforts*
- *Mass revival meetings*
- *Some of those healing events where everyone seems to get involved*

- *Jesus and apostles feeding thousands of people by blessing the food and making it grow.*

I know several of those are religious and there are many others we could address. The point is that when some act that goes against carnal nature occurs and other follow, the resonance level expanded and soon, reality is modified.

# Faith

Jesus called this level quality of resonance faith. Interpreted as a religious characteristic, this faith, or high frequency and high quality resonance is certainly enhanced by strong connection with God. In fact, attempts at trying to establish a very high vibrational level without guidance of God is dangerous. There are many influencers in our universe and not all will try to improve one's level of TRUE awareness. If you noticed from the self actualization chapter, I have three levels attainable above what could be considered Self Actualization.

- *Need of helping others [Self Actualization]*
- *Selfless love [to the exclusion of self]*
- *Understanding of the universe*
- *Understanding outside the universe [Heaven/s]*

Just getting to self actualization is a feat that is all but impossible for most. Even when we reach the level, it is fleeting as basic survival needs keep creeping back in.

## Selfless Love

If one were to describe Jesus, God Incarnate, it would be selfless love. He knew he would be tortured and ridiculed and kill in a horrible way and he not only accepted his fate, he relished its conclusion to show his love. I know others have given up there lives for their children etc. but this level is well above that as most of the people Jesus died for hated him and tried to do terrible things to him all along. If you could just gain a tiny bit of selfless love for small times during a day, you would gain substantial control over your universe.

## Understanding the Universe

Some groups call this the third eye. Being able to see the true universe is not the same as seeing your environment and being at peace with it. That is still self actualization level. This is power by self denial. Remember that faith of a grain of muster-seed can move mountains that Jesus tried to instill. This is that characterization. While is sounds like it would be fantastic to be able to modify universal "law", the huge responsibility comes with it that no one can handle. This is the most dangerous type of control over one's self imaginable as outside influencers WILL bend your will without God. Whatever you could change, it most certainly will be for the bad without close interaction with God. Do not try this one EVERY without close communication with God. The Holy Spirit portion of God worked with the apostles in bringing people back to life, and making wine from water, and all of the things associated with this level which worked out because God was closely attached to them as they ventured into this vibrational state.

## Understanding Heaven

While the Holy Spirit portion of God can help guide us on universal understanding, he becomes paramount in one very important thing, transfer of consciousness to another universe. We can mathematically identify heaven, understand its existence, even travel through it to assess time travel, but living there REQUIRES something identified in many ancient texts as the Light. Without the light, no one can live in heaven. It simply would not make sense to them. Satan and his crones lost the light and had to stay here after dying. That separation is nasty.

## Frequency and Power

The more we allow ourselves to be pulled from the Carnal world, the higher our ethereal dimensional vibration frequency. By increasing our vibrational levels and focusing on what truly is important to provide a level of resonance. We can CONTROL our universe and that of those associated with us.

Let's go back to the fast traveler for an instant. When the traveler speeds up, his entropy is reduced as his mass expands and his potential energy is converted to kinetic [or magnetic] energy associated with traveling through all the atomic clouds in the universe. Both of these reduce entropy and the traveler's consciousness is INCREASED to compensate. He has no sense of time passing and it is believed his awareness of self would also be greatly reduced.

Think of the universe sort of swelling and contracting to the affects of entropy changes. How the changes occur and what we can do with them are not the subject of this book, instead, I want to introduce to you the critical elements of our universe in a different way than you are used to. Its hard enough just looking at things like, time travel, turning into light, sensing time sideways [lateral time viewing], and vibrational transformation of universal elements [how the universe changes as vibrational nodes and resonances interact, expand, and change---regardless of time]. All these things are somewhat controlled by this common collective and by you as you move away from CARNAL thought.

# Atomic Fusion

Let's reexamine Atomic Fusion in regard to Anthropic or conscious controlled reality and vibrational resonance. The law of entropy forces the extinction of large atoms into tiny quasi particles. Today, scientists tell us that **90 percent of the universe is made up of dark matter** or matter that has no apparent mass. Think of this part as the entropic or entropy controlled universe. Try as they may, particles simply cannot form without some outside force. Then this atomic fusion came along. Gluons appeared and started pulling atoms together and building massive ones. Each time a new one was made, massive levels of energy was released. **Here on Earth, 90 percent of the matter is not "Dark Matter".** Why is that? Shouldn't there be an equal distribution of the dark matter stuff? The answer is that there are high levels of consciousness on Earth and we run our reality. While most of the controlling is done in a haphazard way, we can take control is we want to and if we disregard the Carnal characteristics of life that keep us from using our consciousness.

*According to the dictionary, a nuclear force (or nucleon-nucleon interaction or residual strong force) is the force between two or more nucleons. It is responsible for binding of protons and neutrons into atomic nuclei.*

If you think what I am writing is crazy, just try to understand what this nuclear force is. Most scientists have no idea where it came from or what it is, they simply know that atoms are held together by this magic and they want to split it apart to see what happens.

## Nuclear Force is Particle Resonance

Just like other resonance behaviors, Nuclear force gets stronger when particles are larger/ higher frequency. This is because at higher frequencies, the difference in energy associated with resonating a dimensional entity gets more significant to our universe. If we resonate something that vibrates at 100MHz, for instance, we can gain a substantially greater amount of energy than resonating a 1 Kilohertz thing. Higher frequencies and larger particles always affect our universe more. In life, the same is true. High vibrational consciousness affects our universe more than low carnal consciousness. People create massive atom splitters to break the Nuclear force and make subatomic particles of all types. To release or use the energy available from high frequency resonance of structural components such as atoms, simply eliminate the feed of its resonance and it will break down on its own. That is not as simple as it sounds. Because energy is sustained by the other 2-dimensional dynamos, if their resonances are at high frequencies, the particles will tend to resonate at higher frequencies as well. That can be a good thing.

## Lead into Gold

Let's say you wanted some gold and only had lower frequency materials. Simply enhance your "consciousness resonance" or the resonance of the operational dynamo, or the frequency of the structural dynamo directly and gold appears. Here's the problem; this must be done without averting your conscious to the Carnal side and actually want the gold for yourself.

## Water into Wine

Let's say you want to turn water into a higher vibration associated with wine, adjusting your own vibrational resonance will force reactions in the structural and operational dynamos and the water will change. The Bible instruction to the disciples and their subsequent manufacture of wine was in no way impossible. It just takes a different type of concentration [something called faith].

## Stick Into Snake

Let's say you wanted to turn a stick into a snake-like structure just like Jannes, Jambres and Moses did in Egypt so long ago, simply change the vibrational characteristics and the structure will change. I know you have heard this done by the Egyptian magicians [Jannes and Jambres], by Aaron and by his brother Moses and thought it was impossible, but it was not.

## Move a Mountain

Let's say you wanted to move a mountain from one location to another, as was indicated by Jesus, God incarnate, simply change your resonance and force the other 2 dynamos to react by changing theirs causing both structural changes and motion/operational changes. I know all this seems to be too easy. You are probably saying to yourself, *"Why didn't I start doing this long ago?"*---remember you have to separate yourself from the desire for showing people you can do it or you cannot. You need to not try to gain something in the Carnal life, or you cannot.

## Water Walking

Remember that Peter was able to walk on water until he "realized he was doing something he believed to be impossible and ----immediately it BECAME impossible.

## Using a Cyclotron to Increase Frequency

These cyclotrons increase the speed of the particles and stop them quickly to try to transfer operational energy into increased structural energy and it works---sort of. Many times, instead of making new stable, larger particles operating with higher frequency resonance, the particles split apart making a higher operational frequency instead. As the atoms are degenerated, there is a fear that the energy created during the separation of atomic particles could cause disaster. If the opposite happens, a fusion disaster like a nuclear bomb or other scary things could easily take place. One of these cyclotrons in Europe generates so much energy that there were concerns that they might create something called a black hole. One can think of these black hole things as close to the absolute pure gravity where mass is dissolved in the structure. This is like the Soul of the ethereal dynamo taking over control of the Carnal existence, or magnetism almost eliminating the electric potential of the operational dynamo.

If you think I'm crazy, you had better stay away from the guys trying to explain just what a black hole is. If one of these black hole things happened, its resonance would completely change our reality. There would be no more matter in the immediate area and our consciousness could not establish a reality. Electro-magnetics could not establish light so it would be nasty.

Instead of smashing atoms in a dangerous cyclotron Non-Fusion manipulation is a better choice as was described above. During ancient times, some of this Fusion and Disassociation was being done. Sometimes this was accomplished by additional help from the spirit of our creator adding his own vibrational content but other times vibrations were artificially enhanced by will and/or by use of physical components such as crystals. There are so many accounts of ancient people affecting the environment by electromagnetic vibrations by striking crystals or by vibrating of particle structures that there is little doubt that modification of the physical environment was a normal process.

On the simplest terms, we do it all the time. We burn wood, for instance, by simply increasing its frequency with heat until, it changes its characteristics. We generate Radio Waves by simply vibrating a crystal to a high frequency to change potential energy of a battery into a much more changed energy level. We vibrate energy stored in a battery to allow us to build electromagnetic devices that are used fro just about everything. All I'm talking about here is increasing the frequencies being generated and resonating dimensions with those frequencies. Let's see how they did stuff in the old days.

# Vibrational Alchemy

I know this sounds like more fantasy, but people who could change matter by affecting the vibrational patterns in the olden days were known as alchemists. **Moses would have been called an alchemist** and the fabled Merlin would have had that occupation. There is a substantial amount of written history presented around this whole capability. Some of the reactions are covered in this set of books but right now let's concentrate on alchemists.

The alchemists were real and powerful, at least to some level in our ancient past. They used strange tools for their "Craft". One of the tools of these guys was what is typically called the <u>Alchemic tone</u>. [Hopefully the tone or vibration part is starting to sound familiar.]

### The Alchemic Tone

What I mean by Alchemic tone is that it is becoming more and more obvious by the current studies of the unified particle theorists that tone is the answer to the mysteries of the sub-atomic particle. As I previously presented, the defining part of matter does not come down to the 115 different types of atoms, but comes down to the vibrating frequency of even smaller particles. In fact, the particles don't really exist without the vibration. Instead of using particle structures, scientists have now mapped out various components, elements, and systems by wavelength [association with electro-magnetism], and frequency [association with what becomes gravitation]. We're not just talking about radio waves here. If you can produce the frequency components of a substance, you can modify that substance. This modification could mean levitation of a material or even making gold out of another substance. I know this sounds like hocus-pocus, but people are doing this today and they did it in ancient times.

## Enoch

Let's see what the book of Enoch had to say about knowledge concerning alchemy. [As a note- the authority of the book of Enoch is addressed in our current Bible and was most likely excluded because the texts were not available to the clergy that were trying to establish the "canon" books of the Bible.]

***Enoch 64:1-****"A commandment has gone out from the Lord that all that dwell on the Earth shall be destroyed; for they know every secret of the angels, every oppressive and secret power of the devils, and every power of sorcery. They know how silver is produced from the dust of the Earth."* [Clearly alchemy is addressed in this verse. It would be a neat trick to make silver from dirt wouldn't it?]

## Modern Alchemy

Don't think that you will go out and make a tone generator that outputs the 8.5 Exahertz [8,500,000,000,000,000,000 vibrations per second] required to produce Silicon. It can't be produced with existing electronics due to the finite speed of electrons that we can currently pass over any semiconductor material. This type of oscillation could possibly have been manufactured by some means in the very ancient past. This may account for many of the very ancient developments we tend to disbelieve including antigravity, alchemy, and melting stones. There was a reason that the alchemist from the ancient days was feared and worshiped. He could, quite possibly, have been able to make gold out of lead; lift giant stones; and even melt blocks.

## Cinnabar On A Stick

Many of the ancient Alchemists used a rod with a crystal on one end. There is a growing possibility that the crystal structure of these devices was what is called cinnabar. Whether ancient Indians used a large crystal tied to the end of a stick or not, I cannot say, but we do know that crystalline matrices have a characteristic piezo-electricity or vibration after being struck or compressed. Some of the vibrations that can be produced are at very, very, very high frequencies. It is reasonable to believe that this high frequency characteristic would have been most noticeable in crystals of very heavy materials--- let's say cinnabar. I know you are wondering where I got that from. Well, crystalline Mercury [HgS] was well known and mined during ancient times. The name is derived from the

Eastern Indian word for "dragons' blood". Ancient Indian texts are filled with descriptions of some crystals that allowed their ancient vehicles to fly. The crystals were cinnabar along with something called serpent slough. People are going wild trying to find intrinsic properties of Mercury. Let's look at what the very ancient collection of history books called "Manhabharata" had to say.

*Indra and saw thousands of vimanas [flying vehicles] invented by the Gods lying at rest" The ships were-12 cubits in circumference, had 4 wheels, rose in air, and as they flew, a charge of mercury caused roaring flames to shoot out.*

*Another place it says, "Place a mixture of lode-stone, mercury, mica, and serpent-slough on the north and crystals in the center of the engine."*

*Still another passage says, "Place one type of "mani" in sulfuric acid and another type place with magnetite, mica, mercury and, [of course,] serpent-slough. All five crystals should be equipped with wires passing through glass tubes. Wires should be placed from the center in all directions, then a triple wheel will set the revolving motion and the two glass balls inside will turn and increase speed, rubbing each other. The resulting friction generates 100 degree power.*

*Still another passage says, "Place the mercury engine with its iron heating apparatus below. By means of the power latent in the mercury which sets the driving whirlwind in motion, a man sitting inside may travel great distance in a marvelous manner."*

*Here's another that talks about an engine, "Four strong mercury containers must be built in the interior structure, heated by the controlled fire."*

*Still more details are provided, "The vimanas develop "thunder power" from the mercury."*

Like the cinnabar, the serpent-slough stuff is also crystalline so that is going to be even harder to understand. I've never seen Serpent-slough, so I' directing this to the work of vibrating cinnabar/mercury.

### Understanding Cinnabar

There is a slight problem in that the output of many of the experiments is a deadly gas. Breathing the stuff kills the experimenters, but at least they found out that cinnabar is piezoelectric. There are about 500 piezoelectric materials known today. The most widely used is Quartz but Cinnabar is quickly making a name for itself, just like it had done in ancient Greece.

### Greek Cinnabar

In his book "On Stones", Theophrastus of Eresus (371-286 B.C.), a student of Aristotle, described a method to recover mercury from cinnabar by mechanical energy. The metal was obtained from native cinnabar after rubbing it in a brass mortar with a brass pestle in the presence of vinegar. Of course the vinegar and brass was a battery, so they were increasing the vibrational characteristics by forcing electrons through the material. While this doesn't give insight into the vibrational characteristics of Cinnabar, it does show that the crystals were very common. One might believe that a chunk of cinnabar crystal was tied to a stick and the stick was struck to initiate a vibrational sequence that could have done weird stuff. Maybe it could change a person into a newt. We'll look at crystals on a stick shortly. First, let me try to establish a baseline of vibration so to speak.

# Common Material Frequencies

Next are a couple of tables show the actual or theoretical frequency and wavelength standards of common elements known today along with other attributes of other characteristics of our universe. The material frequencies have been derived from the various groups investigating vibration reaction of structures/atoms. How would you like some particles vibrating at 60 exahertz? That vibration causes Gold, as you can see from the list following. Have the right frequency or resonate the environment around a substance and one can make the material you want. Notice that most frequencies do not form matter, at least structures with mass  Even the smallest physical component [BOSON] must vibrate fairly fast [300 MHz] so one would think that if you wanted to modify particles, you had better have a source that can vibrate very, very fast. One thing to note as you look at the tables; vibrating frequencies that create the element we call Meitnerium can even vibrate faster to produce the limits of the Structural dimension or what we call pure magnetism. Some call it a black hole. It is known that the event horizon of a black hole can take dark matter and convert it into much higher frequency "visible matter" and we can do the same thing if we resonate at a high enough frequency.

## Chart of Particle Vibrations

| Name or characteristic | Maximum Wavelength [meters] | Highest Frequency [Hertz] |
|---|---|---|
| Aether [??] | *$1 \times 10^{+10}$ | $<30 \times 10^{-3}$ |
| Fermion [part mass] | *$1 \times 10^{+4}$ | $30 \times 10^{3}$ |
| Boson [smallest mass] | *$1 \times 10^{-0}$ | $30 \times 10^{7}$ |
| Baryon [electron] | *$1 \times 10^{-3}$ | $30 \times 10^{10}$ |
| Hydrogen/1 | $1 \times 10^{-9}$ | $30 \times 10^{16}$ |
| Berylium/9 | $1 \times 10^{-10}$ | $30 \times 10^{17}$ |
| Silicon/28 | $3.5 \times 10^{-11}$ | $8.5 \times 10^{18}$ |
| Zirconium/91 | $1 \times 10^{-11}$ | $30 \times 10^{18}$ |
| Gold/197 | $5 \times 10^{-12}$ | $60 \times 10^{18}$ |
| Meitnerium/270 | $3.7 \times 10^{-12}$ | $27 \times 10^{19}$ |
| Straight Gravity | smaller | higher |

For completeness, here are the things we believe we know about the other 2 dynamos with respect to vibrations.

## Chart of Electro-Magnetic Vibrations

| Name or characteristic | Maximum Wavelength [meters] | Highest Frequency [Hertz] |
|---|---|---|
| Electricity | $5 \times 10^{10}$ | $<30 \times 10^{-3}$ |
| Brain function | $5 \times 10^{7}$ | $6 \times 10^{0}$ to $10^{1}$ |
| Human hearing | $1 \times 10^{4}$ | $20 \times 10^{3}$ |
| VHF [radio] | $1 \times 10^{0}$ | $30 \times 10^{7}$ |
| UHF [radio] | $1 \times 10^{-1}$ | $30 \times 10^{8}$ |
| SHF [radio] | $1 \times 10^{-2}$ | $30 \times 10^{9}$ |
| EHF [radio] | $1 \times 10^{-3}$ | $30 \times 10^{10}$ |
| Microwaves | $2.5 \times 10^{-4}$ | $12 \times 10^{12}$ |
| Infrared [light] | $1 \times 10^{-6}$ | $30 \times 10^{13}$ |
| Visible light | $4 \times 10^{-7}$ | $75 \times 10^{13}$ |
| X-rays | $1 \times 10^{-8}$ | $30 \times 10^{15}$ |
| Gamma Rays | $1 \times 10^{-9}$ | $30 \times 10^{16}$ |
| Magnetism | lower | higher |

** it is highly likely that brain function frequencies are simply catalyst for much higher frequencies actually used by our brains to store thoughts and images.

The thing we know about electromagnetic frequencies is that less input energy is required for a particular action the higher the frequency of the action. It becomes easier to attain a purer, higher quality resonance. Model life the same way and look at the next chart.

## Chart of Ethereal Vibrations

| Name or characteristic | Maximum Wavelength [meters] | Highest Frequency [Hertz] |
|---|---|---|
| Molecular Interaction | $5 \times 10^{10}$ | $<30 \times 10^{-3}$ |
| Unaware Life | $1 \times 10^4$ | $30 \times 10^3$ |
| Life Awareness | $1 \times 10^0$ | $30 \times 10^7$ |
| Survival | $1 \times 10^{-3}$ | $30 \times 10^{10}$ |
| Sex | $1 \times 10^{-9}$ | $30 \times 10^{16}$ |
| Need for Companionship | $1 \times 10^{-10}$ | $30 \times 10^{17}$ |
| Need to help others [Self Actualized] | $3.5 \times 10^{-11}$ | $8.5 \times 10^{18}$ |
| Selfless Love | $5 \times 10^{-12}$ | $60 \times 10^{18}$ |
| Universal Understanding | $3.7 \times 10^{-12}$ | $27 \times 10^{19}$ |
| Insight into the External World | smaller | Higher |

My list is somewhat different that those called chakra levels by the Buddhists, but there is a similarity that cannot be ignored. No matter how you sense it, increasing your vibrational resonance frequency, increases your power over your environment. At the very high frequencies, there is almost no need for the environment. Eliminating the lower frequency elements allows you to sustain the higher levels longer and better.

# Beat Frequencies

That brings us to crystals and something called "Beat Frequencies". You have probably heard about crystals having some magical power and dismissed it as some type of belief destined to go along with astrology and extracts of poppy seeds. The Tamashii model of atomic structure and this whole concept of vibrating particles may give credence to the notion that crystals hold magic. If you start with a crystal of a homogeneous material that is locked in a covalent lattice structure, it will tend to vibrate at a very specific frequency when excited and the vibrations will continue for some time due to the resonance of the crystalline substrate. In other words, a crystal could cause a continuing vibration. A secondary vibration from a sound cue or other stimulus could very well produce a "beat frequency". This means the two frequencies will "ADD" in a special way. If you add a 1KHz signal with a 1.1KHz signal, they can beat together and make a 2.1 KHz frequency that will beat again and make a 3KHz signal that can Beat with the 2KHz signal to make 5 KHz and so on. If you start with really high frequencies, the beat frequencies get huge. If you want the QUALITY to increase to sense the higher frequencies, simply dampen the lower frequencies or ignore them. Get a high enough frequency and a different material could be created.

## Caution

Don't discount the magic crystal thing, but don't go out and get a crystal to make you feel better either. It probably will just sit there and do nothing for you. That will have the opposite affect as ethereal vibration levels require focus outside the normal environment.

## Beat frequencies of Life

Just like the previous example, if 2 consciousnesses get together and somehow raise their frequencies, just being near another will make the frequencies even higher as they beat together.

There have been hundreds of experiments on this theme and all are spooky. People can solve problems faster if many people are trying to solve the problem EVEN IF THEY ARE NOT COMMUNICATING WITH EACH OTHER.

## Bad Beat Frequencies

This whole get in a group trying to expand awareness thing has just as bad of an effect. If a mob builds and lowers the consciousness frequencies with debased ideas, and attitudes, the entire group will be lowered EVEN IF THERE IS NO VERBAL CONTACT BETWEEN EACH OF THE MOB MEMBERS. This is because beat frequencies go both ways. A 1 KHz signal and a 1.1 KHz signal ALSO make a 0.1 KHz "SUBTRACTED" frequency. We see the beat frequencies in electro-magnetics all the time and we can easily sense beat frequencies of sound pressure. Believe me when I tell you, the same happens with consciousness.

**Get with the wrong people and you can greatly limit any greatness you can have. Get near the right people and your world will be opened up.**

*Its not magic it vibrations and resonance.*

Ancient humans discovered that vibrations, beat frequencies and power were all related. Open your mind to possibilities that ancient humans could do marvelous things with crystals and other particles that we are only now beginning to understand, but there is the warning.

Because they were talented in these exotic ways, the people of that time quit listening to the creator God and started having self-centered thinking. They were empowered by their own science and did not understand how the ethereal dimension had to react to allow for true happiness and peace. Many ancient texts talk about how horrible the science was during that time and how the people had no idea about what that science was doing to them.

That being discounted, there were many discoveries associated with changing vibrational resonances of materials and electro-magnetics. Let's look at some of them. As we do try to compare these physical vibration enhancing things to what might have been possible if they enhanced their ethereal vibrations.

# Vibrating in History

As it turns out, making vibrating potential energy to make useable electricity is one of the easier things to accept from our distant past. It could very well be that the Arc of the Covenant, Moses magical staff, the Egyptian Dendera tubes, turning sticks to snakes, levitating blocks, and other things that keeps showing up in our history are true. What was happening with all of these varying types of "mystery things" is that vibrations produced by some controlling medium converted matter just like the photon changing from visible light to the deadly gamma rays. If one can control matter in some minor way, another thing that could be increased is life itself by causing the DNA or structure of life to be modified. As we confront the possibilities of matter being controlled by vibration, many of the things that seemed to be fantasy or some religious exaggeration were actually the truth. You wondered how Peter walked on the water and how Elisha's bones brought a boy back to life or how the Red Sea was raised from its river bed. Some simply say God did it. It's magic. Well God doesn't work like that. He created this world with a set of physical laws and the things he or any of his creations do in this world work within those laws. We are only now beginning to understand what some of those laws really are. The most important is the law of vibrational matter. The bringing people back to life one will have to wait until we get into the last of the 10-dimensions so don't worry about that one just yet.

# Vibration Staff

Our Bible talks about vibrating staffs doing all types of unusual things. The staffs somehow collected the electricity and used it for required tasks. One of the tasks seemed to be making tones to change atomic structure. One of the obvious things that occur from this is levitation. In ancient texts, we are told of sounds or vibrations coming from one of these staffs, which eventually made heavy stones weightless. All of it sounds like some crazy talk, I know, but the alternative is to discount hundreds of ancient documents, drawings, verbal histories, and even claims from the Bible which rely on some magical quality of several magical staffs during ancient times.

## Priests of Hike and Thoth

Thoth was the first of a long line of Egyptian magicians called the Priests of Hike and he had one of these staffs. In the Emerald Texts it was written that Thoth left Atlantis and survived the flood to teach the Egyptians. He was probably one of the ANAK and he began teaching the

Egyptians about writing and all types of what we think of as magic and science today. He was revered as a god, was the thirteenth ruler of Egypt, and was probably responsible for teaching man "magic tricks" and making some of the machines that could perform the magic that are discussed below.

### The Mattah Staff

Thoth may have introduced the magic staff [also called the Mattah or UAS by the Egyptians]. It is, by far, the most used and best known "magic device". The power in many of these things was, most likely, awesome. One of the "tricks" it was known for was the ability to "open a lake or river". Not only did Moses have great power when using this device, but other historical records show the strength of the "Staff" when used by other magicians before Moses had his opportunity. Some of the better known magicians that were able to perform amazing feats, aided by a staff, besides Thoth, are the Hike priests, Bimater, Mises, Jajamankh, Elijah, Elisha, Moses, Aaron, Joshua, and possibly a even a non-fictitious Merlin. All of these people used "staffs" effectively to do things that are thought to be physically impossible to do today. As a note of interest, Moses staff was made from wood, but the common method for building a UAS in ancient Egypt was to allow a bull's penis to dry in a "Stick-like" shape. It was considered to bring power to its owner.

### Staff Levitated Stones

One of the more important priest tricks was to move the huge stones by levitation. This was discussed before in the section dealing with construction techniques. According to ancient texts, this was done by using one of the magic staffs to produce some kind of tones that would allow the stone to be lifted and carried through the air for a short distance. I have no idea if the stick had a cinnabar crystal on it or some other generator, but we are certainly told the sticks could transform matter just like electricity does only in a more exotic way. From the looks of the many huge stones used as building material, this levitating thing was almost common knowledge by the "Magicians" after the flood. The ancient descriptions indicating that the staff would cause some type of tones that somehow raise objects off the ground goes along with the ancient Indian texts which discuss levitation and flying machines had something to do with sound.

## Staff Levitated Water

The staffs didn't just levitate rocks. They also levitated water. The first story written about levitating water was during the time of king Seneferu of Egypt. He had lost a trinket in a lake accidentally and asked one of the Hike Priests named Jajamankh to get it back. The magician simply took his staff, held it up and the water immediately rose up so that he could walk to the center and retrieve the trinket. Upon lowering his staff, the water filled its normal position. This happened about 2000BC. Here are some of the "magic staff" stories that have survived. Note the Mandaean version of Moses' escape from the Egyptians. That story provides us with a possibility that another river was raised, but during that act, a mistake occurred that costs hundreds of lives. Although different from the Biblical text, the Mandaean similarities establish consistency in historical record.

***Orphic Hymn to Bacchus*** - *According to this Babylonian work, the sorcerer named Bimater had a rod with which he could work miracles very similar to the one that Moses had. The "miracles" including parting waters of the river Orantes, the river Hydastus, and the Red Sea. [This same man also struck a rock and produced water and had his rod turn into a serpent.]*

*2 Kings 2:8- And Elijah took his mantle, and wrapped it together, and smote the waters, and they were divided hither and thither, so that they two went over on dry ground. [Elijah parted the waters.]*

*2 Kings 2:13-14- He [Elisha] took up also the mantle of Elijah that fell from him, and went back, and stood by the bank of Jordan; And he took the mantle of Elijah that fell from him, and smote the waters, and said, Where is the LORD God of Elijah? And when he also had smitten the waters, they parted hither and thither: and Elisha went over. [Elisha parted the waters.]*

*Josephus- Moses had thus addressed himself to God, he smote the sea with his rod, which parted asunder at the stroke, and receiving those waters into itself, left the ground dry, as a road and a place of flight for the Hebrews. [Moses parted the waters according to secular history.]*

*Exodus 14:21- And Moses stretched out his hand over the sea; and the LORD caused the sea to go back by a strong east wind all that night, and made the sea dry land, and the waters were divided. And the children of Israel went into the midst of the sea upon the dry ground: and the waters were a wall unto them on their right hand, and on* their left. *[Moses parted the waters according to Biblical history.]*

*Mandaean tradition [Iran]- Most Jews worshiped Ruha [Lilith?] and knew nothing of the "Light" or the "children of the Light". The Jews tried to escape the Egyptians. Upon their escape, one of them named Musa [Moses] had a staff, and knowledge of secret names. The staff had been given to him by Ruha [Lilith?] and opened into two parts. When they reached the Sea of Suf [Red Sea], Musa took the rod and struck the water and uttered names and the water became solid like ground. The Egyptian followed and the magician Para Malka used his Marghna [staff] and also made dry land. The magician continued across the sea while the Egyptians came across. When the magician reached the far shore, the sea turned back into water and all behind him was drowned. [Malka and Musa both parted the waters.]*

*Moses was the most well known Jewish Magician, but he was far from the only one. Jannes and Jambres were also Jewish and well known for their practices, but they were not doing their trade for God. These types of people were known as sorcerers and the followers of Moses were told to STAY AWAY from their magic.*

# Moses Staff

Moses was most likely a Hike Priest and was aware of many of the magical incantations common in that day. This did not only include levitating water out of its bank or movement of huge blocks, but also included turning the Mattah [staff] into various animals. Some drawings are shown below of the staff. The ancient rulers and gods to show royalty did not hold up the Mattah; these people held it because it helped protect them. Certainly, the later rulers held the staff because the earlier rulers had done it, without understanding why, but initially it was a powerful tool. Notice that on top of the stick there is some type of long thing [possibly a cinnabar crystal].

# Lotapes, Jannes, & Jambres

Just because a magician was Jewish did not mean that the magicians was necessarily good. Two Jewish magicians [sometimes called sorcerers] did everything that Moses had done to mock his magic. Their names were Jannes and Jambres and they were famous for their magic. They made snakes, frogs and everything else. They probably used a Mattah just like Moses did when he initiated the plagues on Egypt.

***Jannes and Jambres**- And in the presence of the King [of Egypt], he [Jannes] opposed Moses and his brother Aaron by doing everything they had done.*

***Targun [Exodus 1:15]**- Then Jannes and Jambres, the chief wizards, spoke up and said to Pharaoh---*

*Gospel of Nicodemus- They were servants of Pharaoh, Jannes and Jambres, and they also did signs not a few of which Moses did, and the Egyptians held them as gods. [This shows that the Jewish magicians Jannes and Jambres were not only magicians, but they were also very highly placed in the Royal family.]*

***Historia Naturalis**- There is another magical group deriving from Moses, Jannes, Lotapes, and the Jews, but many thousands of years after Zoroaster----.*

***Numerius**- Next are Jannes and Jambres, Egyptian sacred scribes, men judged to be inferior to none in magic, when the Jews were expelled from Egypt, they were chosen by the people of Egypt to stand up to Moses, the leader of the Jews, and a man of most powerful prayer.*

*Chronicles of Moses-* And after Moses and Aaron left, Pharaoh sent and called to Balaam the magician and Jannes and Jambres his son the sorcerers.

## Other "Staff" Magic

In the hands of Moses, the staff became an awesome weapon. Not only could Moses' Rod aid in levitating water, but it could also perform other miracles including the following: [Although God probably helped Moses a little, it is very apparent from other texts that the "magic staffs" of the day could do many things.

*Turning into a snake-* The Sumerians, Egyptian magicians [Jannes and Jambres], and Moses were all able to accomplish this feat while holding a Crystal headed staff.

*Manufacturing water when the ground was struck-* The Sumerians and Moses accomplished this feat while holding a staff.

*Turning water into blood-* The Egyptians [Jannes and Jambres] and Moses both were able to do this while holding a staff.

*Turning bad water into clean drinking water-* Only Moses reportedly did this feat while holding a staff.

*Bringing huge quantities of Frogs-*The Egyptians [Jannes and Jambres] and Moses were able to do this while holding a staff.

While holding his staff, he was also able to bring locust, darkness, and many other miracles including death. Normal Egyptian magicians could not do these other miracles no matter how they twisted their sticks. The reason Moses could do more than the other magicians was that Moses' rod was aided by the power of God, according to historical record.

Here are some more magical things that were attributed to the "staff" and written down in historical records. Some of the magic was done by Moses and some was done by other magicians of the day.

***Egyptian Emerald Tablet-*** *Then I [Thoth] raised my staff and directed a ray of vibration striking them in their tracks.*

***Exodus 7:20-*** *And Moses lifted up the rod, and smote the waters that were in the river and all the waters that were in the river were turned to blood.*

***On Baptism 9-*** *Similarly, water was healed of its bitterness and changed into fresh drinkable water by the staff of Moses.*

***Exodus 7:22-*** *And the magicians of Egypt did so [turned water into blood] with their enchantments:*

***Exodus 8:6-*** *And Aaron stretched out his hand over the waters of Egypt; and the frogs came up, and covered the land of Egypt.*

***Exodus 8:7-*** *And the [Egyptian] magicians did so with their enchantments, and brought up frogs upon the land of Egypt.*

***Josephus-*** *Moses put the rod down upon the ground, and commanded it to turn itself into a serpent. It obeyed him, and went all round, and devoured the rods of the*

*Egyptians, which seemed to be dragons, until it had consumed them all. It then returned to its own form.*

Below is a table just a small sampling of the "known" magical things accomplished while holding a staff. Moses was the king of the magical staff. This does not include general forms of levitation called out in many texts because specific people were not identified.

| | Moses [Jew] | Mises [Sumerian] | Malka & Musa | Elijah & Elisha | Jajamankh-Egypt | Bimater [Babylon] | Thoth [Atlantean] | Jannes & Jambres [Egypt] |
|---|---|---|---|---|---|---|---|---|
| Raise Water Of A River | ■ | ■ | | | | ■ | | |
| Water from Rock | ■ | ■ | ■ | | | ■ | | |
| Turn bad water to good | ■ | | | ■ | | | | |
| Turn Rod To Serpent | ■ | | | | | | ■ | ■ |
| Call Frogs | ■ | | | | | | | ■ |
| Turn water to blood | ■ | | | | | | | ■ |
| Used As Weapon | ■ | | | | | | ■ | |
| Call Locust | ■ | | | | | | | |
| Make it dark | ■ | | | | | | | |
| Heal Serpent bite | ■ | | | | | | | |

# Levitation and the Djed

You didn't need a stick to change matter in the olden days. A djed, like the one shown, was just as good. As levitation has been connected with changing the vibrational structure of matter, I think we get more of an appreciation of how one can change his environment easily. Back in Egypt, we find some more references to these Djed things. The picture-word "Djed" is depicted as the symbol "medu", which is usually interpreted as making the single sound "SAH". From hieroglyphs, one can sense that even the gods of Egypt worshipped these things. Whatever this vibrating thing was, it could change matter and control the physical environment. From the text following we can assume that the Djed-device may have also had something to do with making an environment that allowed for levitation, just like that described in earlier examples of levitation. Actually, it is not a difficult leap to believe the DJED had something to do with flying in general. These excerpts below come from the Pyramid Text "utterance 539. Let's read the verses to see what I mean.

*[1303b]* *A Djed making the SAH sound will allow him to levitate and become empty on the way to the Blue Void. [Somehow, the Djed and levitation are linked. The blue void could have been a number of things and this was no ordinary SAH sound, so don't start chanting.]*

*[1304c]* *(A Djed making the SAH sound), will allow the sight of Pepi I to be as "the Opener of the paths" [The Djed also seems to produce radiation of some kind.]*

*[1309a]* *(A Djed making the SAH sound), will insure that the two shoulders of Pepi I will be as Illusion. [The Djed seems to be associated with invisibility.]*

*[1313a]* *(A Djed making the SAH sound), will insure that the buttocks of Pepi I will be as the Right-levitator together with the Left-levitator. [Somehow the Pepi I could steer with his butt? I'm not going to explain this one. Just watch where you sit when you are at Egyptian ruins--- especially if your butt cheeks are not coordinated.]*

*[1315a]* *(A Djed making the SAH sound), will insure that the soles of Pepi I's feet will be as two correctly-positioned levitators. [Possibly it is not talking about the person Pepi, but some kind of ship with flames coming out of its base which would be similar to the feet. Then the butt makes more sense.]*

*[1316a]* *(A Djed making the SAH sound), will insure that Pepi I will be one who is on the way to be a god, or a son of a god. [Seems to be referring to going into space.]*

*[1317a] (Whenever the Djed is properly used), Pepi I will be reincarnated for the god-Star. [Again space or heaven are referenced.]*

*[1318c] (A Djed making the SAH sound), will insure that Magic will be in the body of Pepi I. [Assuming Pepi was a person, the Djed had something to do with producing Magic.]*

*[1320c] (A Djed making the SAH sound), will insure that there will be Second Sight and the Youthfulness of a young person. [This seems to indicate that the Djed somehow controlled aging.]*

*[1324] It is not Pepi I who speaks to the gods; it is Magic, which speaks to the gods. (A Djed making the SAH sound) will insure that Pepi I reaches the lower realm of Magic. [The Djed was strongly connected to Magic.]*

*[1325c] (Whenever the Djed) is heard, the gods will provide him a seat in the levitating-god-Star. [The Djed somehow made the user able to ride in the "levitating god-star" I have no idea if the seat has butt controls so just get that image out of your mind.]*

*[1327a] The gods will accept the control of Pepi I on the way to the Blue Void, because he has already gone to the house of Second Sight, which is in the Calm Void Current, so that the nature of his consciousness will be true to the Soul. [Even control of (the spaceship?) was possible.]*

## The Djed Again

Utterance 239, talks about Djed medu just like the preceding one. Here is what the texts had to say.

*[243a.]* *A Djed produces the sound "sah", so that the White Crown Crossing Power will engage in Levitation. It is because it has become aware of the Mahatmic Eye. [The Mahatmic Eye is associated with second sight while levitation is clearly raising an object off the ground. Both seem to be associated with the Djed. Other than that, I don't know what the verse is trying to say.]*

I don't know exactly what this thing was, but it was something amazing and don't let people simply tell you it was nothing more than a symbol of Osiris's backbone. That is absurd. From the previous verses we can see that in addition to the possible connection with extending life, the Djed was thought to have many other capabilities including levitation, magic, second sight and controlling someone's butt. No wonder it was so revered.

# Arc Power Anomaly

As we travel away from Egypt, we find more clues of changing the environment [presumably by means of vibrational control]. There is evidence to suggest that Moses might have siphoned electricity to help power the "Arc-of-the-Covenant" by making its physical characteristics that which allowed for it reception. There seems to be a sign of a build up of electricity inside a large storage device. How massive amounts of high voltage electricity could have been generated in the Arc has always been a question. Some simply attribute it to God. They say the physical laws he designed for this world were simply eliminated, but they does not sound like the creator God. He would simply design Physics around what he wanted and he wanted the Arc to be able to become electrified with NORMAL physics. So long as the Jews worshipped him enough, the Arc was powerful. As they drifted back into the Carnal nature and didn't focus or expand their awareness, the Arc lost power. We can think of our control over our environment in the same way. As we drift away from our carnal elements [Sex, Self and survival], we can begin to change our environment. If we can push our consciousness to a higher plane, we can affect more of it. Here are some of the characteristics of the Arc of the Covenant.

## Moses Builds It

After Moses constructed the magical Arc, I would assume that it was "charged" by the output energy from the pyramid and the "focusing "staff" he carried with him always. Possibly, he simply modified the materials inside the arc to produce ultra-high voltage Electrical charges by means of focusing various vibrational electromagnetic waves from his staff. At the right combination of vibrations, the "essence" of the Arc became a weapon and it spewed destruction. I know this isn't the story you heard previously, but the evidence suggests this is a more responsible discussion and it shows that vibrational modification of particles was well know during that time and the Arc may have shown us the extent..

This small device, covered in gold [the great electrical conductor], not only was said to have housed the commandments of God and other sacred items, but it also killed many people, according to many references. In order to get close to it, people had to remove their shoes. After a while, it got so dangerous that the Jewish people didn't even want it near, even though it had been responsible for several of their wins during battles against their enemies and it had made it possible for the Jews to cross the Jordan River. Along the top of the arc were golden cherubim, but it was certainly more than a box. After all, it was designed and made by one of the highest ranking members of the royal family in Egypt and, presumably, he had been schooled in the arts of magic just like many of the other royal personnel. One of the "Magical" capabilities was, most likely, the ability to produce electricity.

**God Made Fire**

If you don't accept the electricity theory, there could be other explanations for the Arc's miraculous power, but just saying, "God focused energy through it", is probably not the way God has worked in the past nor is it probably the way he helped the Jews with this device. His grand order includes physical laws that he works through. We've seen it millions of times everyday. Possibly the only two times he did some type of creation that was separated from known natural physics was at the creation of Adam and the embodiment of Jesus, and those occurrences probably were governed by some natural law as well. We just don't know what it was at this particular time.

In addition to sending out electric charges, the device could cause rivers to dry up and kill anyone closer than 2000 cubits [about half a mile]. There were indications that lightning came out of it and death came to those who touched it. It got so bad, that whenever others captured the arc, it was quickly returned to the Jewish people, and the Jews didn't even want it near them because they feared its power. This thing was like a huge electrical capacitor, which possibly stored thousands of volts of electricity.

***Joshua 3:3-4-*** *And they commanded the people, saying, When ye see the ark of the covenant, and the priests bearing it, then ye shall remove from your place, and go after it. Yet there shall be a space between you and it, about two thousand cubits by measure: come not near unto it, that ye may know the way by which ye must go*

***Joshua 3:15-*** *And as they that bare the ark were come unto Jordan, and the feet of the priests that bare the ark were dipped in the brim of the water, the waters which came down from above stood and rose up upon an heap [The arc also moved water away just like Moses' rod had done earlier.]*

I think it is pretty clear that energy was collected by the Arc just like we collect electro-,magnetic energy in the various systems we have today.----By tuning to the specific vibration of the transmission. During ancient times, people were able to tune into the vibrational components of our universe.

We could go on and on about great ancient sorcerers and magicians. I use the word great only as a reference to magic capability, not as some people to necessarily be respected. Let's look at what the "Book of Secrets" had to say. This book was found among the Essene libraries known as the Dead Sea Scrolls.

# Book of Secrets

The "Book of Secrets" found with the Dead Sea Scrolls provides a strong warning about the use of "secrets of God". I think you will be able to tell from the texts that the secret was the manipulation of matter and electromagnetics by changing the vibrational resonance of the environment.

The book simply says that if we use this "magic", the same thing will happen to mankind that had happened before if we can't change our own resonating frequency. The earth would be destroyed again. This destruction would not be by direct intervention of God, but because we, as humans, don't understand what we are doing as we manipulate "Nature". Of the secret elements indicated in the text, it seems that the "manipulation of creation" is the worst. This probably references the genetic manipulation and transmutation or one material into another [Alchemy]. Here are the major elements of what has been pieced together of the "Book of Secrets". Judge for yourself. If it makes you fearful, you probably read it correctly. (The bracketed portions have been added by me in areas where missing information was noted.) The highlighted elements are just comments.

*Those who would penetrate the origins of knowledge, along with those who hold fast to the wonderful mysteries (of magic;) [This is talking about all humans that practice the secrets of "magic". The concept of penetrating the ORIGINS of knowledge lets us know that this is very ancient science being discussed in these verses.]*

--*Those who walk in simplicity as well as those who are devious in every activity of the deeds of humanity; those with a stiff neck [self centered], and all the mass of the Gentiles, [All humans including those that were not pure blood Adamic humans were not left out of this warning. Even if one thinks he is "walking in Simplicity" he won't escape the ravages that are caused by use of the "SECRETS". Note that the Gentiles hold a special place of badness as it indicates the "ENTIRE MASS" of gentiles. By the way, a gentile is any mixed breed, not pure blooded Adamic, This includes ALL people alive today alive today as no pureblooded Adamics are around anymore.]*

--*With (this I beseech your attention. All of the) secrets of sin (and magic were known) but they- [The "they" here is talking about the very ancient humans from before the flood.]*

-*did not know the secret of the way things are nor did they understand the things of old. [This section indicates that no one knew the ramifications of meddling with nature even before the worldwide flood. It is saying no one understood the REAL affects of using things like genetic mutation and Alchemy. Changing those things without changing the vibrational resonance of our consciousnesses DESTABLIZES our universe. ]*

*They did not know what would come upon them, so they did not rescue themselves without the secret of the way things are. [Magic did not warn or save anyone from the flood. This thing called the "SECRET OF THE WAY THINGS ARE" is a secret that was not known to anyone. They, evidently did not understand the delicate resonance of our universal dimensions.]*

*It is true that all the peoples reject evil, yet it advances in all of them. Who wants his money to be stolen by a wicked man, but where is the people that has not robbed the wealth of another? [According to this passage, no human can escape evil. Even if he thinks that he is doing good, it can "AND WILL" change to evil. Certainly one can reject the physical/carnal nature for a time, but it is not possible to sustain that level of resonance.]*

*What shall we call man who will call no one on earth wise or righteous? It is not a human possession (to act on wisdom.) [This seems to be saying that occasionally, humans can get to a vibrational level that allows them to see our reality for what it is, but, too often, these insights are short lived.]*

*It is not (possible because) wisdom is hidden except for the wisdom of cunning evil, and the schemes of Belial (who modified creation,) a thing that ought never to be done again, except by the command of his Maker. [This is the important part. There is an conscious entity that keeps our understanding of the truth from continuing and it always wins. Belial, in this passage is another name fro Satan and his dominions.]*

*(God controls) every secret, and he limits every deed and what (magic that is known by) the Gentiles, for He created them and their deeds also [The magic that is done by mixed breed humans is occasionally allowed by God. None is accomplished without his knowing.]*

*Consider the soothsayers, those teachers of sin and (magic. Do not regard) your foolishness, for the vision is sealed up from you, and you have not properly understood the eternal mysteries. [Just in case you didn't get it the first time, it is stated again. Manipulation of nature and Magic cannot be understood by man. It is foolish attempt these things without enlightening ourselves.---I know we are using genetic engineering today to help many things, but this is saying the apparent help will not last.]*

*You have not become wise in understanding (my secrets); for you have not properly understood the origin of Wisdom. [Here is all one has to do to be able to correctly use "magic". In order to understand how to manipulate nature in a good way, you must make yourself at one with wisdom.]*

It seems that many of the sciences we "sort of" rely on every day are inappropriate and will eventually get us into serious problems. Today there is a general belief that AIDS was manufactured for another purpose, but turned into a killing animal or germ. Every time we go forward, we get sidestepped with a different catastrophe.

All this sounds too ominous to think about and possibly, you think that not changing water into wine is a more reasonable direction to follow, but, the overriding concern in this ancient text is that our consciousness level must be expanded if we are to do these things. Removing ourselves from carnal characteristics, keeps demonic pressures away, keeps self indulgence from shadowing the effects, and allows our reality to change without causing the reaction to be problematic in another area of reality. The last part is the important part here. If one does change an event, there should be no attempt at continuing in the effort as ANY vibrational frequency and resonance is short lived without a strong enhancement by this Holy Spirit" that was given to the early Apostles to allow for massive modifications over a sustained amount of time. Any time the spirit left them, they would quickly drift back into their normal Carnal nature. Even after they have been schooled for years by Jesus, they still could not sustain the level of increase vibration for long periods. That being said, modern experiments are showing that people in large groups affect reality more than single or small groups.

# People in Groups

As the Jews all worshipped in one accord, the Arc gained in power. As multiple of Gentiles "practiced" magic, they gained in inappropriate power. There are examples everywhere and these help us understand one key point about changing reality. The more trying to accomplish a single task the easier it becomes. I'm not talking about more people working, I simply saying that getting in a group of people that think like you will modify the environment to be what your group needs.

Many experiments are going on today describing this seemingly unbelievable characteristic of humans that Einstein talked about. If your conscious mind affected your reality, many concentrating on the same things, change the environment more. These Social Knowledge experiments show that minor changes in reality can be directly noted by just having an arbitrary group becoming likeminded. Other experiments test mob rule with hate being the controlling element. Others, who would not normally become violent, become violent and have almost no understanding about why it happened. The aggregate consciousness began controlling the vibrational tone of the people nearby. It changes reality.

Monks take this the other way. If they stay away from those thoughts and try to eliminate the sex, self, survival vibrations, they change the "Normal" vibrational tone of the environment and allow a resonance that is higher frequency and has more control of their reality. There are reports of levitation of objects including themselves, complete peace with the barest of necessities, and other seemingly physics violating phenomena. All of this stuff seems to be emanating from vibrations. Sometimes the vibrations are mental and not measurable at the present time and other times, pressure way, audio tones can be heard during the modification. Monks chanting and humming, the Djed making the SAH sound, Joshua having all the Jews shout to destroy the wall of Jericho, and Incan and Babylonian stories about levitating stones with a sound describe how sounds are heard when the environment is changed. It seems that even measurable vibrations, somehow affect reality. Just as music allows us to drift away from the carnal thoughts, the vibrations may not be the direct source of the changes, but sounds that reduce the carnal vibrations, certainly can help us become sensitive to higher vibrations.

# Tesla Vibrates

That is where Nikoli Tesla might give us a hand. Nikoli Tesla reinvented what we call AC current or vibrational current not too long ago. Just to make things interesting, let me tell you something odd about alternating current. There is no direct exchange of electrons through a material so there is no electro-negative potential that can cause work to be done. It makes absolutely no sense that having electricity go back and forth in a wire could make motors run or electronics "electronate". After saying that, let me tell you the  alternating current does work. It works by, momentarily, energizing a structure such that it can make atoms get larger and then reducing the size of the atoms. This vibration cause the magnetic fields of the structure to materials to build and the collapse of the magnetic fields allow work to be accomplished. It is just like magic, but we take it for granted every day. The master of this unbelievable secret, Tesla, indicated that its discovery magically came to him.

Not only did Tesla discover alternating current, but he also discovered that the ground could carry alternating Electricity, magically.

**Besides Telsa, no one in recent history has been able to use the ground to carry electricity.**

The ground has moisture in it and metals of all types. When electric current is pushed through the ground, the ground simply heats up and no electricity is transferred. In Telsa's configuration, no electric wires were needed to carry electricity as most people believe must be required today. He demonstrated to many this new capability and, reportedly, lit large quantities of electric lamps, which were placed at long distances away from his generator. He, somehow, used the earth as the conductor and lit them without wires. The problem with his newfound distribution method was that no one could control who was using the energy, therefore, no Earth transfer systems were ever commercially produced. No one would gain a profit and Tesla's method died with him.

Tesla's typical experimentation involved substantial high voltage discharges as shown on the next page, but he really wanted to do something special for everyone in the world and, it seems that he was able to change reality with the huge electrical energy things.

He wanted to make free electricity for everyone. Unfortunately, he could not get the backing needed to further his experiments in this area. He had received financial backing from J. Pierpont Morgan of $150,000 to build a radio transmitter for signaling Europe, but later he wrote that he had much larger plans. With the first portion of the money, he obtained 200 acres of land at Shoreham, Long Island and built an enormous tower 187 feet tall topped with a 55 ton, 68 foot metal dome. This was to be the beginnings of an electrical distribution unit that used the earth as a conductor. He was unable to complete his experiment, probably because there would be no money in it for J.P. Morgan who withdrew funding. In fact, everyone abandoned the conquest; so today, we must pay for our electricity. I know that you are probably saying to yourself that conducting electricity through the earth is impossible, but now we know it is not impossible, nor is it that simple.

## A New Automobile

Later Telsa was said to have used his free air transmission of electricity to power an automobile. After showing how the car could operate without the nee for gasoline, these experiments were also defunded and abandoned before Tesla died. The 1931 Pierce-Arrow shown next was supposedly the type converted to run on the unknown energy source. Possibly, he had found a way to create electricity by manipulation of the vibrational components in free air.

**Possibly The Engine Used Earth Conduction**

**Confirmed Earth Conduction**

It really wasn't until 1976 that Dr. Andrija Puharich was able to point out that Tesla's power transmission system could not be explained by the laws of classical electrodynamics, but, rather, in terms of relativistic transformations in high energy fields. It seems that when an electron encounters a positron, the two particles would annihilate each other. Because energy can't be destroyed, the particles are transformed into an electromagnetic wave. Transversely, if a huge electromagnetic wave is manufactured, electrons and positrons of equal quantity must be manufactured. This recombination can be done at a remote site. Let me put this whole thing in the frequency domain and associate it with Vibrational matter.

All one would have to do is to somehow make the electricity go out of phase with this universe and it would no longer exist in this reality. At a destination, simply relink the electricity to this reality by converting the phase back to what it should be. It is believed that Tesla was able to change his environment consciously or unconsciously.

Ordinary electrical currents do not travel far through the earth. Dirt has a high resistance to electricity and quickly turns currents into heat energy that would be wasted. With this "pair- production" method, electricity can be moved from one point to another without really having to push the physical particle through the earth. The transmitting source would create a strong field, and a particle would only be created at the receiver by means of some catalyst. Without the catalyst, there would be no loss of energy and no heat would be generated. Put another way, the earth could sort of store the electricity until needed.

Evidently, the ground current distribution of electricity during the ancient times was used for many years. Caves full of paintings have no carbon on the ceilings so they had a different form of light than fire. Electroplating seemed to be common 5 thousand years ago and farther back. The production of Aluminum and Titanium that both require substantial electricity were well known. On and on we find that the ancient people knew about this whole vibrating thing. Unfortunately, they didn't seem to understand the vibration of the consciousness.

Nikoli Tesla used the power of vibrating high voltage electricity to change reality. Others just think about peaceful thoughts to change their reality.

## Keely Vibrates

I'm not getting into the exploits of John Keely in this book, but he, evidently affected his reality in some way as well. John preceded Tesla in his experimentation about vibrating masses changing things. The unusual thing to note about John Keely is that he seemed to have no real idea how he was affecting reality. He wrote tons of articles about the Aether and what he called the Aetheric vibration causing all types of energy releases, and he demonstrated many exciting machines that worked on the principle, but when he was not in the room and actually touching whoever was running the experiment, the experiments would fail. Keely died not understanding what he had accomplished. Possibly Tesla also had only a vague idea of what he was doing. We will never know the limits of his understanding because his last thesis on what magnetism really was taken or destroyed when his mysterious death occurred. He had levitated, made electricity go through the ground, and even participated in experiments in space time distortions. He had tapped into his consciousness and controlled his reality.

The easiest thing to sense with respect to matter manipulation by your consciousness is something we call levitation.

# Levitation

If your consciousness vibrates fast enough, you can levitate just like Peter, Elijah, and Elisha walking [or levitating] across the water in the Biblical texts. I know Jesus, God incarnate, walked on water a lot as well, but he is a special case.

## Levitating People

This next section doesn't seem possible, but the evidence is well documented and witnessed, in many cases, by large groups. Many people have had the ability to lift their own bodies into the air over the years as indicated below. No, I'm not talking about David Copperfield here. Some of them lost their lives because it was considered to be an act of the devil, but some survived and floated around. If this list seems impressive, just think of the numbers we would know about if levitation wasn't considered demonic in the old days. Arthur Herbert Thurston wrote the "Physical Phenomena of Mysticism" about many instances of levitation. Some of those listed below came from his investigations and other sources are otherwise identified.

**St. Bernardino Realino-** *He initially had a glow come from his body before he rose in the air in 1608 according to testimony of a nobleman of the time.*

**St. Adolphus Ligori-** *He was supposedly raised into the air in front of his entire congregation in 1777 according to the same work.*

**St. Joseph-** *He would cry out and be lifted up without regard into trees and above his congregation, according to Catholic traditions.*

**Father Francis Suarez-** *He reportedly began to glow with a blinding light before being lifted, according to a catholic monk.*

**Palladius-***This Roman historian wrote about a child levitating as seen by his own eyes in the 4th century.*

**Moslems-***In the twelfth century an Iranian was noted for his frequent flights to the treetops.*

**France-***The French historian Louis Jacolliot witnessed 2 levitations in India and recorded them in "Occult Science of India".*

**Daniel Douglas Home-** *He was probably the most recognized levitator of the Victorian age and was seen by many to float on several occasions during the 1870's.*

**Joseph Gianvill-** *He wrote about levitation by demons in his book "Saducismus Triumphatus" in 1681.*

***Oliver Gilbert Leroy-*** *He wrote in "La Levitation" stories of 230 different Catholic saints with the ability to levitate.*

***Dr. A. Imbert Gourbeyre-*** *He wrote "La Stigmatisation". It contained over 200 Levitation events as well.*

***Colin Evans-*** *He was well known for his ability to levitate himself and many people saw him accomplish the task in the 1930s. Sometimes he was photographed as he floated.*

These people could change their reality.

# Human Combustion

Hopefully you are sensing that these things we normally think of as things change into completely different things all the time and we simply accept it without looking at the fundamental assertion that things we take for granted as things are nothing more than particular frequencies of vibration.

Let's assume our body increases its temperature to protect us from infection. Don't think of it as heat killing bacteria but low frequency photons. If the body can do dramatic changes in its internal vibrational structure for that, it can do even more  amazing changes. Sometime these changes are less than gratifying and many times, accidental "Beat frequencies" can take hold of the physical components of our bodies and make a mess of things.

## Human Vibrational Combustion

Here is a thing to consider here. Just like electromagnetics gets hotter as it vibrates faster, particles get hotter as they vibrate faster. If a person is in the wrong environment, this can become deadly. We are told mass can't be created and then we are told that photons sometime have mass and it was created simply by vibrating a mass fast enough. Hopefully you are seeing how all of this stuff is sort of melding together. There is no difference between mass, light, heat, electricity or consciousness. It's all the same thing [vibrating nothingness]. Speaking of converting body temperature to photons, all these ancient tests about the aura of an angel or person becoming visible to others, should be no surprise. If you can vibrate you body to make it kill bacteria, certainly you could make yourself glow. It's not magic or even odd. I'm sure that if your eye registered slightly lower photonic vibrations, you would see everyone glowing. It certainly would be neat to make yourself into a light bulb and the bacteria would all die in the process. Wait just a minute. That sounds like spontaneous human combustion.

Lightning may not be the worst way to die. What about death by fire? Most people have heard about spontaneous human combustion and immediately ignored it due to its absurdity even though there have been very few explanations presented for some very bizarre occurrences. Some had even seen pictures like the one on to the right that showed the affects of an unexplained phenomenon, but were unmoved. Unfortunately, for Jacqueline Fitzsimons, in 1985 it became too real and there were many witnesses of its reality. She just went to college one day, told her friend she wasn't feeling well and that her back felt extremely hot. All of a sudden, her shirt caught on fire and she screamed for help. Many came to her rescue and put the flames out, but not before her hair was a blaze. After 15 days in intensive care, she died. History and science ignore what has been happening in the past. It is now a little harder to ignore. People can catch on fire by some internal means, so if you get heartburn carry antacids and possibly a fire extinguisher. The remains of Dr. John Irving Bently is shown on the previous page. On the following page is a short list of the more notable instances that have been determined to be spontaneous human combustion in just the past 50 years. This is not an isolated case.

- *1951,July 1-Mary Reeser burst into flames. Only the corner of the room was incinerated.-Florida*
- *1950's- A girl burst in flames. Her boyfriend couldn't put flames out.-London*
- *1953, May- Esther Dulin burst into flames in a chair. Only she and the chair were consumed.-LA USA*
- *1957,May- Anna Martin burst into flames at home.Pa. USA*
- *1959, Jan. -Jack Larber burst into flames in a hospital-San Francisco*
- *1959, Dec.-Billy Peterson burst into flames in a car. The seat was generally undamaged-Michigan*
- *1960,Nov.-5 men burned up in a house, carbon monoxide in lungs suggested breathing moments before death & high rate burn-Kentucky*
- *1964-Fiery death of a man was reportedly like an exploding person-London*

- *1964, Nov.-Helen Conway's torso was found charred at home.-Pa, USA*
- *1964, Oct.-Olga Worth Stephen burst into flames in her car-Texas*
- *1966, Mar. -John Greeley was consumed in his car.-Land's End*
- *1966, Mar. -George Turner was found incinerated in a ditch.-Chester*
- *1966, Mar. -Willem Bruik burst into flames in a car.-Holland*
- *1966, Dec.-Dr. John Irving Bentley was incinerated at home. His leg was unaffected .[shown]-Pa USA*
- *1967, Sept- Robert Bailey burst into flames from his stomach.-London*
- *pre-1976-An Englishman was incinerated in his truck, but the gas tank was unaffected- England*
- *pre-1976- A man on the street burst into flames.-London*
- *pre-1976-A Contractor burst into flames while waving to employees.- England*
- *1980, Jan. 6- Mr. Blackwood was completely incinerated but rubber on his walker was not burned.-Wales*
- *1980, Oct.- Jenna Winchester burst into flames in her car.-Florida*
- *1982, Aug.-A woman burst into flames on the street.-Chicago*
- *1982, Sept.-Jean Saffin burst into flames in her kitchen. Her father remained helpless.-London*
- *1985- Jouqueline Simmons burst into flames at school.*

- *1989-A 27 year old engineer burst into flames. His stomach and belly were carbonized.- Hungary*
- *1997, March 24-John Oconner burst into flames at home. Only his head, upper torso and feet remained.- Ireland*
- *1998, Aug.-Agnes Philips burst into flames in her car. Flames came from her chest cavity.-Australia*
- *2003-Alexei Rusnac's head was burnt down to the size of and orange. His clothes were not burned "reported in Ananova"- Rumania*

What the list shows is that "Spontaneous Human Combustion" occurrences are all over the place, and people spontaneously combust just about every year. Some years more incidences seem to happen than others and the following general information concerning the symptoms of the mystery are described below.

- *Although it doesn't appear that way from my short list, almost eighty percent of the victims are female.*

- *Most of the victims were overweight and/or alcoholics.*

- *The body is very badly burned, but the room the body is found is in pretty much intact except for a fine layer of soot.*

- *Yellow, foul smelling oil usually surrounds the body.*

- *The torso, including the chest, abdomen and hips tend to be totally consumed, sparing portions of the extremities and the head - the clothing can also be intact.*

- *Generally, victims were on their own. Generally, no shouts or screams could be heard. This was not the case with Jacqueline.*

- *The victim had usually been drinking heavily prior to the death, but that isn't always the case.*

The list doesn't help very much as we try to find a cure for this affliction, but hopefully you are beginning to believe that there are things about our human bodies that we have little knowledge about and I don't only mean that one day we may blow up.

## Spontaneous Feline Combustion

Of course, humans are the only animals that should worry about this phenomenon. No one will probably know about most of the explosions, but Peppi the cat's flames were seen quite clearly. The day was the 28th of November 1986. He was an 8-year-old house cat and just sleeping on a chair at the Anmer Lodge in Stanmore London. Witnesses said that there was a terrific bang followed by a flash going a few feet in the air. Peppi was enveloped in a blue flame similar to other spontaneous combustion victims and essentially disintegrated. Peppi was another victim and still there is no answer to the mystery.

## Combustion Discussion

As with most everything else, I have a vibrational theory about this strange anomaly. All the other explanations seem to fall away from one incident or another, but there may be a broad theory that allows for and explains how this terrible thing could happen. The Tamashii atomic model helps explain what may be happening. With this general concept in mind let me say that if the right electromagnetic conditions surrounded the unfortunate "combustion" victims, some of the molecules in their bodies could have easily been converted into some type of combustible material. That doesn't mean that when you get heartburn you should go for a fire extinguisher, but it does mean that there is a logical explanation for things that may initially appear to be odd. New researchers may be getting closer to the answers of these exploding people by making bowling balls fly in the air on their own. I know that sounds like an impossible thing to happen, but that doesn't mean that it didn't happen. We will find later that a Canadian named John Hutchison is causing many "impossible things" to happen by using extremely high frequency signal bombardments. After many successes and more failures, he can tell you that it is extremely difficult to get the right combination of fields to cause changes in materials, but when they change, the effect is dramatic and you don't need to use ATOMIC FISION or FUSION or anything else that could be bad. Human combustion is one of the dramatic changes that can occur from a changing FREQUENCY. There might, however, be a reason why many of the instances involved alcohol. The possible reason is that alcohol causes desensitization of nerves in our body which allows for misfired electrical signals. Information paths are, then, disrupted more by the chemical imbalance at the nerve endings. While the misfired signals would probably not cause the reaction on their own, they may help in the modification of the internal molecular makeup that results in a "self induced" explosion. Just like extreme anger seems to bring our bodies to an extremely heated condition, this action probably requires many individually unimportant events to occur simultaneously.

I know this explanation doesn't help us control, or know the specifics about why each of the events occurred, but it does provide a link between spontaneous human combustion and reality. As electro-magnetic beat frequencies are funneled or even accidentally manufactured, very strange anomalies like levitation, body combustion, healing someone with the mind, and even out of body visions may result. Each of these anomalies have had many, many, witnessed occurrences. Just because we don't quite understand, the physics doesn't mean that there is not a real physical law that allows for these strange things to be part of our normal life. For a short time, conversion of materials from noncombustible to highly flammable occurred inside Jacqueline Simmons and Peppi the cat so don't say it is too far fetched to happen.

# Ed Leedskalnin

Let me give you another example of someone who could change what we call reality. If you can change matter by vibration, you could change its weight, just like the ancient one had done. The guy, shown on the right, immigrated to the United States from Latvia in 1923 and quickly contracted the deadly Tuberculosis disease. Somehow, he cured himself. That wasn't the amazing thing to consider here. He also "harnessed gravity itself" it would seem. He picked up 1100 tons of coral rock, carved the stones, and built a castle "alone", and with no help whatsoever. In 1952 this apparent magician died, but before he passed away he told many about his ability to control gravity. After his death, a strange machine was found at his castle. As shown on the following page, the machine had many magnets embedded in rock and others that were positioned by turning a crank on the top. While the portion of the machine seems to be incomplete, we can image this machine to be somewhat similar to the no-energy machines being produced around the world today, but this one, somehow affected gravity. As the machine was turned, it would have generated some frequency associated with the times that the rotor magnets passed near the embedded ones.

If the generated vibrational components could have been high enough, the vary essence of the nearby materials he was trying to affect would have changed. Evidently, the boulders he moved turned into something quite different than what they are today so that he could easily place them in their locations. They, evidently, lost most of their mass or gravity while under the influence of Ed's machine and his consciousness.

Magnetic Stator    Strange array of critically place magnets

Mr. Leedskalnin insisted that all matter was not made of atoms but, instead, consisted of individual magnets and it is the movement of these magnets within materials [Vibrational element] and through space that produces measurable phenomena, i.e., magnetism and electricity. He also claimed to know how the Great Pyramid was built, and to prove it he moved a 30-ton and other monolithic blocks of coral to build his castle. He inferred that out-of-phase gravity waves can be created in such a manner to neutralize in-phase gravity waves. While most of my discussions is about changing particles by changing the vibrational component remember that the gravity component appears to be a perpendicularly phased vibrational component that works in tandem with the normal vibrational component associated with the characteristics of matter. Ed understood what ancient humans had known during the very ancient times and we are just now relearning.

No one really knows what Ed did, but you can be sure it had something to do with vibrating magnetic fields converting matter.

# John Hutchison

Luckily, Mr. Leedskalnin wasn't the last man to rediscover levitation, disintegration and free energy by vibration assisted with consciousness. We cannot talk about modern levitation without discussing John Hutchison. He currently lives in Canada and has been conducting some amazing, well witnessed, experiments  since 1979. He will soon find out what produces what we call the Hutchison effect once he decides that he affects his experiments. Right now let me just list the elements known so far. He has demonstrated all of them in the presence of a strange field of electromagnetic waves.

- *Objects became temporarily invisible*
- *Heavy objects can levitate [even a bowling ball]*
- *Things can pass through each other*
- *As they pass through each other there is no apparent change in either component physical characteristics.*
- *Sometimes metals can become like jelly*
- *Sometimes metals melt without heating*

Hopefully your interest has peaked. The inventor uses multiple radio transmitters and high voltage Tesla coils concentrated at a single location to produce "something" and when it is produced, the above things occur to objects in this "Field". If you noticed a link between the multiple radio transmitters and the descriptions of the "Tamashii model of atomic structure" and the use of vibrations to transform elements, I think you are on the right track here. By changing the electro-magnetic resonance in an area, we can more easily affect things by conscious endeavors.

 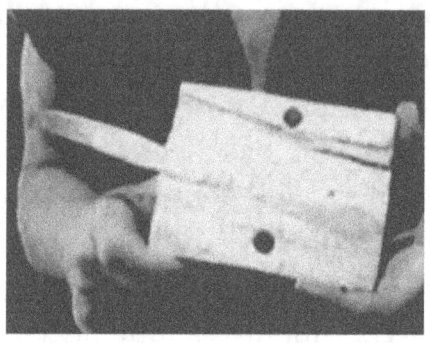

John Hutchison has accidentally found the correct vibrations for particular elements to make them appear invisible to one another so that one can pass through another. The picture above to the right shows how a simple butter knife was pushed through a piece of metal without anyone's help. Now cut in half, the butter knife has become part of the metal. He also has made objects appear to be gravitationally invisible to the earth to allow levitation. During some of his experiments, vibrations mutated the materials to appear to be melted or jelly-like. [See the metal bar above to the left.] Sometimes metals would appear to melt but surrounding wooden objects would not get hot as only the metal structure changed in the field. The effect was as if the objects were changing from this universe to another. The stuff John was finding out was incredible, but someone got scared. In 2006, the Canadian Government went up to John's home and confiscated all of his vibration altering equipment. While "vibrationally" we may be able to define and even modify a particle, if we are to understand the universe, we must expand our awareness to the 10-dimensions that make up the vibrational entities.

# Invisibility and Alchemy

People have been trying for years to replicate ancient capabilities in invisibility, alchemy and many other things by building large magnetic vibrational fields. Sometimes we have gotten close. One time might have been called the Philadelphia experiment.

### Philadelphia Experiment

If we could better understand the Tamashii Atomic model, Einstein's unified particle theory, Tesla's Ground current phenomenon, the Acoustic Laws of Keely, the Hutchison Effect, and the ANTHROPIC PRINCIPLE that have been touched on throughout this work, maybe we could better understand what might have happened in Philadelphia on board the USS Eldridge during World War II as shown next.

Most of the reports of that experiment can't be verified and initially seemed fanciful. The reports of people becoming fused with the bulkheads, and the ship disappearing and reappearing at a different location all seem more like science fiction, but maybe there is more to the stories that we can begin to understand. I'm not getting into this area as many already have heard about the outcome of this experiment except to say that people should not have been inserted into the experiment. People were supposedly found merged with bulkheads and the survivors went mad. Rather than suppressing these experiments, they should be heralded. If we can understand these oddball physical anomalies, maybe we can better understand something we call Alchemy.

## Modern Alchemy

This section is on successful transmutational experiments done in the 20th century. I know alchemy is a bad word and instills an image of sorcery, but in the 20th Century, it is almost commonplace to transmutate elements into other elements. The following list describes the successes that have been reported. Today most are done with the "Atom Smashers" so it is still too expensive to make gold for most of us. As can be seen from the table below, some are beginning to use high voltage and frequency methods which go along with the Tamashii model. These methods don't take up nearly as must space as an atomic accelerator.

| Date | Researcher | Method | Initial Mat'l | # | End | # |
|---|---|---|---|---|---|---|
| 1920s | Franz Tausend | A | Mercury | 80 | Gold | 79 |
| 1923 | A. Miethe | B | Mercury | 80 | Gold | 79 |
| 1924 | Hantaro Nagaoka | B | Mercury | 80 | Gold | 79 |
| 1924 | Smits & Karssen | B | Lead | 82 | Mercury | 80 |
| 1927 | Walter Russell | B | Oxygen | 8 | Nitrogen | 7 |
| 1935 | Lord Rutherford | C | Nitrogen | 7 | Oxygen | 8 |
| 1936 | Various | C | Platinum | 78 | Gold | 79 |
| 1938 | Hahn & Strassman | C | Uranium | 92 | Ba + Kr | 56+36 |
| 1939 | Various | C | Thorium | 90 | Radium | 88 |
| 1960s | Jnana & Caro | A | Lead | 82 | Gold | 79 |
| 1972 | Soviet Physicists | C | Lead | 82 | Gold | 79 |
| 1980 | Glenn Seaborg, | C | Lead | 82 | Gold | 79 |
| 1986 | Various | C | Mercury | 80 | Gold | 79 |

With this list, there is a strong echo of the "Book of Secrets" warning. According to that ancient work, God warned that-- *"Man can change elements just like he can change animals, but he cannot understand the outcome of his actions."* By the way; the A means-Harmonic Alchemy [not really frequency based]—B-means High voltage/high frequency distillate—C-means Nuclear Accelerator Bombardment. Notice that the most converted element is Gold. I guess there is a flare for sorcery in our modern day scientists.

142

# Beyond

Some of you are wondering why I have gone over these bizarre quasi-scientific experiments using anomalous methodologies and identified questionable applications of possible uses of vibrational knowledge from thousands of years ago. The answer is simple. No one else seems to be providing this information to those who might be able to use the information the most. That group is our children. They may be able to better understand these "phenomenon" if they can build on the previous experimentation. Hopefully, you are considering the possibility that the ancient texts and the other evidence of almost magical capabilities were not lies or fairytales. Our new understanding of matter in general and how things can completely change characteristics by simply changing a vibrational element of a system is now allowing us to accept ancient events that previously separated religious accounts and scientific understanding. Heaven not only is a possibility but a necessity and the probability of angels existing now has become reasonable. Lastly, a photon can be a particle sometimes and a wave [or vibration] sometimes.

I need to go a little further to allow a more complete vision of this new concept of reality. Reality is not stable. Reality is not the same for everyone. Reality can be controlled by changing the vibrations your consciousness, changing electromagnetic vibrations, and by changing vibrations or particles. All three dimensional dynamos build your reality, not just one thing. The one you have the most control over is your consciousness. Part of the consciousness is set by self awareness, sexual desire, and survival. Those can be considered your SELF. Consciously going outside those characteristics in thought, awareness, and action can be dimensionally called your Spirit. Light electromagnetic fields, as the frequency goes higher, energy levels increase and the limit is something we can call pure magnetism. In life, extending our spirit to the limits of vibrations can be called your SOUL.

# Soul and Spirit

The soul is the complete oneness with reality. In the particle world, it is the same thing as complete gravity or the Black hole. At the edges of a black hole Matter is created. At the edges of the collection of souls, LIFE IS CREATED. In the operational dimension dynamo, pure magnetism creates all FORCES.

The Bible tells us that gaining all the riches of the universe is not worth losing your SOUL. You can't really discuss Spirit and soul separately and be very accurate.

The ancients called this spirit thing "the light". Later on, the Bible redefined it [because "Light" sounded like light bulb I suppose and the apostles had never seen one of those.] In fact, the New Testament started talking about 2 things that are beyond life. Those things were called "spirit" and "soul". Some try to tell you that they are the same thing, but the Bible is clear that these are different. The "Spirit dimension" joins life and consciousness together and is the interface to other universes. By the way, transcendalists and all those other people who talk about leaving their bodies are not talking about having their "spirit" leave even if that is what they say. It is a soul which can leave the body as a conscious entity. We know there is a difference between the soul (psyche or consciousness) and the spirit (pneuma). Psyche is the transference of the conscious mind. Let me quickly provide a few Bible verses that generally describe these 2 components of what I call the Ethereal dynamo.

*Hebrews 4:12-* "*For the word of God is quick, and powerful, and sharper than any two-edged sword, piercing even to the dividing asunder of soul and spirit...*"

*John 3:5-6-"When you are reborn, you are born of spirit"*

*1 Thessalonians 5:23-* "*And the very God of peace sanctify you wholly; and I pray God your whole spirit and soul and body be preserved blameless unto the coming of our Lord Jesus Christ."*

# Light

### Satan's Light

As punishment from the Heaven Wars, Satan and his followers were turned human, except the "LIGHT" was taken from them so they could never live in heaven.

*Let's examine this little bit of religious history.*

These men were now called the ANAK. If we go along with what the book is trying to reveal, we can sense that these particular humans could invent things, could sense carnal life to its fullest, but the things that they could not do without the LIGHT was to change their reality. Many ancient texts talk about the struggles that the ANAK had trying to regain this light thing.

### Holy Spirit Light

In the New Testament, Jesus indicated that he was sending down his "Holy Spirit" to allow us to "See the Light". Once the Holy spirit interacted with a person's spirit, he could face death without fear. He could raise people from the dead, he could heal the sick, and he could generally speak for God. The examples of the exploits of those who received the Holy spirit are numerous. God's assurance in the Bible is that the Holy Spirit can help ALL who want him to come. The problem is that the 3 S's keep getting in the way. One must abandon, self, sex, and survival to allow this Holy Spirit entrance to modify the vibrational resonance and allow a better control over one's reality including the reality of what happens after what we call death.

## This Is Not The End

In the next major section of this book, we will see that the going to the light and feeling the warmth and all the rest people tell us from near death experiences may not be the end. Nor is it a one way tunnel. All the warmth stuff, friendliness, and other feelings seem to all be their, but death is a very strange component of life. Let me finish the life part and I will get right on death in a more open minded way. While we are alive, let's also look at how easily the consciousness can leave the body [Alive or Dead].

# Out Of Body Consciousness

Surely you have heard about out of body consciousness! People simply leave their body and float around. I'm not just talking about when there is a near death situation; I'm talking about doing this sort of thing as desired. It is estimated that about 1/4 to 1/3 of the population depending on which study you look at have experienced SOME type of out of body thing. Sometimes it is just for an instant, a feeling that you already did something or know something that will happen, but many of us lose linkage with the COMMON consciousness we depend on. Many times the experimenters are totally aware of their surroundings. They recognize objects including, dead people who can talk to them. These adventures are not identified as being dreamlike, but instead are very REAL feeling.

One type of out-of-body experience is call astral projection. In this method of increasing ones conscious vibrations, people indicate that they sometime prepare several days and focus on a place that they wish to project to and they have various techniques to place themselves in a hypnotic state. Sometimes a simple word or "Mantra" is used to set up a situation. Once the initial conscious level is elevated, they feel like they are flying and they must fight not to fall into a deeper sleep level. And they feel at peace with the universe.

Many times people who have these experiences are changed forever. I think this is something close to the crown chakra level that ancients attest to. They are experiencing modification by Anthropics.

# Carbon 12 Anthropics

I know I keep saying Anthropics but it doesn't mean anything so here is a little example called carbon. Carbon 12 is the building block of life but Carbon 12 doesn't make sense. Nuclear and biological scientists can't figure out how carbon-12 was made in the first place. Even beyond that the heavier massed elements are even more bizarre. Carbon-12 is made from Helium-4 and Beryillium-8. With modifications of chemicals typically being binary in nature, only one atom will be modified in a nuclear incident, but Beryillium-8 is a modification of 2 helium-4s. When we make this substance, it only is beryllium for $1 \times 10^{-17}$ seconds before it turns back into helium-4. It is not enough time to produce the combination. Even today, no non-anthropic explanation of how carbon-12 is made has had any real meat.

*Carbon-12 seems to not be possible, but there Carbon-12 is building people.*

Carbon-12 was made because we needed it to become entities. With that, let's look at Anthropics in a little more detail.

# The Anthropic Principle

The idea that consciousness affects reality is not new, but now it has a sophisticated name "The Anthropic Principle" or the Anthropic Universe Theory". No one likes this idea because it makes the concept of time more difficult, but the whole Quantum Mechanics era has really destroyed our concept of stable time and stable reality, so, we are stuck with it no matter what. The whole "Power of Positive Thinking", "Think and Grow Rich" and all other concepts of the 70s which tried to convince you that how you consciously view reality will affect reality are not only true, they affect your death as well. In the Anthropic World if you have faith of a grain of muster-seed, you can move a mountain, as Jesus said thousands of years ago and you can walk on water as demonstrated by Elijah, Elisha, Peter, and Jesus did so many years ago. With the Anthropic Principle, science and religion can act as a single tool for us to understand God, the universe, and ourselves.

Before we can understand Anthropics, we have to look at quantum mechanics, I'm afraid. Speaking of quantum Mechanics, 2 different experiments have recently been able to transport information associated with a particle and regenerate the particle at a long distance without time delay. For Star Trekies, I said that they are now teleporting things using a quantum engine that eliminates space-time. The Anthropic Principle tells us that WE (all conscious life) CAN affect the physical characteristic of ALL things. Let me give you an example.

### Electron Example

According to many Anthropic Scientists, an observer is very deeply involved in any quantum event. On the simplest terms, an electron cloud surrounds a central core of an Atom. Rather than spinning, it is now known to be in superposition as something we could call a wave-packet. It is located at all locations around the nucleus. This only changes when one wishes to sense it. Finding a location of an electron takes 2 things. One is one must disrupt its field of position in some way and look for results, but the other is that a cognizant viewer must be integrated in the observation or the electron will stay in superposition. It is, at the moment of observation, by a conscious mind that the electron "chooses" one of the possible locations to materialize in, collapsing its wave-packet and becoming a particle for the split second it takes to be hit by the other particle.

Another way of stating this is that electrons are sometimes a wave and sometimes a particle. Our interaction with the universe converts it.

In electro-magnetics finding out that light was sometimes a wave and sometimes a particle has been described form many, many years. It is only now understood that particles act exactly like light.

The logical conclusion of Athropics is that Life can be associated with something that could be considered particle-like [Carnal thought, brain control, muscle reaction] and sometimes wave-like [having the ability to control our destiny and reality]. The wave-like characteristic of Life and Death are the keys to happiness, understanding, reduction of fear, and oneness with our creator.

I know some of you are skeptical about your being able to shape reality with your thoughts and you don't see how that ability can affect your death but hopefully, after you read this book, it will be clearer to you. If you were skeptical about life, I guarantee you will have a somewhat difficult time with the subject of death. Just to be clear let be say that I'm discussing how death fits into a vibrational world where even our conscious thought can change the very framework of our existence. Even after death, there is more to learn. For the most part, I will examine both external characterizations and religious discussions so we can compare details as we go.

### Ascent of the Blessed by Hieronymus Bosch

The 15th century depiction of conscious "spirits" of people passing from this world to another through a huge lighted tunnel was the depiction 600 years ago [see next] and it is still the belief of many. Most people believe death is a one way trip to happiness or doom. To add to the confusion, most sense this lighted tunnel thing during "near death experience" which further intensifies the belief, but it seems there is much more to it. Most of the religions of the world recognize, like the Jewish and Christian teachings, that there is a cyclic nature to life and the various versions of the SOUL are important to establishing our reality.

**Taoist Belief-**In Taoism, there is not exactly an afterlife. While that sounds final, they believe that we are eternal in Tao. The afterlife is, sort of within life itself. People are of the Tao when they are living and when they die, they are the Tao

**Hindu Belief-**Hindu believe in many Gods controlling different segments of life. Hindu have this doctrine of "anatta", which is the notion that individuals don't really possess eternal souls. Instead of eternal souls, we consist of a bundle of habits, memories, sensations, and desires. Together these things make us think we are in a stable, lasting self.

**Buddhist Belief-** Buddhist have a similarity to both Tao and Hindu in that they have doctrines of reincarnation, karma, the notion that the ultimate goal of the life is to escape this thing they called the cycle of death and rebirth. They believe we are bound to the death/rebirth process by desire or craving anything in the world.

**Moslem Belief-** If you kill enough Christians, you will have an eternity of virgins. If you are a woman, you will go to hell.

**Kemetian Belief-**The ancient Egyptians told us - "The forces of darkness were not conquered forever at the beginning of time; instead they surrounded the Earth as serpents poised to attack God. Even with this EVIL, they believed that existence in the "afterlife" is eternal, and one who "measures up" will be like a god, striding forward like the "Lords of Eternity".

**Inca Belief-**The Inca wanted to regain their bodies in the "afterlife" so they mummified dead rulers, children that had been sacrificed, and a group known as the "Cloud people". At some point in time, the "wanderers" could regain their bodies.

**Aztec Belief-** For them there could be no new life without death so the killed a lot of people to insure they could be reborn.

**Greek Belief-**The ancient Greeks believed that at the moment of death, a person obtained a higher level of consciousness.

**Sikhism and Eckankar-**Eckankar focuses on spiritual exercises that enable people to experience something called "the Light and Sound of God." Once one attains the "light", he becomes an Eckankar or "Co-Worker with God".

OK! These are strange religions, but the main theme with respect to anthropic science is that something in entities that are conscious are attached to and help define what we call reality. As someone modifies how he interacts with this group, reality changes. If Carbon 12 was required for the entities, it will be in the reality of those entities. If someone witnesses a tree falling, it will make sound. If no one is around, there is absolutely no need for the tree in the first place so it will not be their. It is sort of written into existence as needed. If we could go to the end of our universe and NOT BE ALIVE, we would not see a star, a galaxy, or even dark matter. Looking back at us there is nothing.

I know this is a strange concept, but it is important for understanding life and how to expand awareness in this life. It is important because it is the only truth that has been able to describe our universe completely.

# Jews

While Jewish people had some ideas about life resonance, they were pretty mixed up about how life ended. I put this section together to establish some position on death in terms many of us already understand. While some of this next set of verses is specific to the various Jewish religions, it seems reasonable to put them together so that we can see transition of thought and compare them to the other ancient teachings from around the world.

**Jewish Sadducees-** These guys were in between Jewish and Zoroastrians who had built up a following in Persia. All of a sudden Heaven hell, and even the final rule of God on earth started to push into the Jewish religion. The unusual thoughts of this sect did not stay long as over 90% of the population, rose up to reject all Persian concepts including resurrection, angels, and spirits. This group of Jewish people had a little bit better idea of life after death but rejected reincarnation, so the Pharisees came along.

**Jewish Pharisees-**A group of the Jews started reading the "Holy Books" again and determined the Sadducees were wrong. This group became known as Pharisees. Unlike the Sadducees, these Jews did believe in reincarnation. They believed the souls of evil men were punished after death. The souls of good men were "removed into other bodies" and they "had power to revive and live again." Still, the Carnal life was the end goal of these Jews. It was not Heaven.

**Jewish Essenes and Pythagoras-** One group started looking to Pythagoras to build a religion. He had done a number of studies into reincarnation and the Greek and Pharisee religion came together and produced the Essenes. They also believed in non-violence to all living creatures. The Essene believed that the soul was both immortal and pre-existent. Oh Boy, pre-existent is much more than reincarnated. Our souls were from the beginning of time. Another thing odd about the Essene is that they had a book about the end of time. Pieces of it have been found and it speaks about what happens when you die. This book sounds identical to the "Revelation" book in the Bible and it was written well before the time of John who wrote the later one. Let me give you an example.

**"The Essene Book of Revelation"-** *And then I looked, and behold, a door was opened in heaven: And a voice which sounded from all sides, like a trumpet, spoke to me: "Come up here, and I will show you the things which must be hereafter." [transmigration] And immediately I was there, in spirit, at the threshold of the open door and I entered through the open door <u>into a sea of blazing</u>*

*light. And in the midst of the blinding ocean of radiance was a throne:* ["Near-death Experience with a bright light and a tunnel" Don't event let someone tell you that seeing a tunnel of light is simply people trying to say what others said. These guys were long ago, before newspapers.]

**Gnostics and Plato-**Instead of going with Pythagoras, one group began combining Jewish beliefs with those of Plato. These were the major believers of "transmigration." Unlike the preexistent soul, this group believed that one could allow your "Soul" to migrate out of your body even before you were dead. I suppose you could say that the Gnostics believed that you could <u>transmigrate often</u> if you wanted to and the <u>Christians and Essene indicated that only special circumstance allowed for it.</u>

**Modern Jewish** –I suppose you can see that Jews are probably really mixed up so they really concentrate on living in the Carnal world but their holy books talk about a different understanding. Here is what their "Zohar" says. *"All souls are subject to revolutions."* [reincarnations] *"Men do not know the way they have been judged in all time."* (Zohar II, 199b) Therefore, Jews would be judged for bad things and have no memory of the things they were judged for in the Afterlife which is just not fair. The "Kether Malkuth" say, *"If the soul is pure, then it shall obtain favor. If it has been defiled, then it shall wander for a time in pain and despair until the days of her purification."* You might wonder, "How can the soul be defiled before birth?" Jewish Rabbis help and say this verse means that the defiled soul wanders down from paradise through many births until the soul regained its purity. My feeling is that Jews really don't know what to believe.

**Ecclesiastes and Reincarnation-***"What has been will be again, what has been done will be done again; there is nothing new under the sun ... Whatever is has already been, and what will be has been before; and God will call the past to account."* Certainly, reincarnation is discussed here, and in the Jewish belief, God would remember all of your past lives as part of the full accounting to see if you went to a good place.

**Jeremiah Confirms Transmigration-***Jeremiah 1:4 says even more about existence before birth. "Then the word of the LORD came unto me, saying, before I formed thee in the belly I knew thee; and before thou camest forth out of the womb I sanctified thee, and I ordained thee a prophet unto the nations." This is talking about life before the womb or what we might call cognizant death between lives if we are reading a book on death.*

**Job 1:20-21** fills in more detail. *"Then Job arose and tore his robe and shaved his head and he fell to the ground and worshipped. And he said, "Naked I came from my mother's womb, and naked I shall return there. Job 19:25 "I know that my Redeemer lives, and that in the end he will stand upon the earth. And after my skin has been destroyed, yet in my flesh I will see God; I myself will see him with my own eyes-I, and not another." [Reincarnation]*

***Ecclesiastes*** *1:8 tells us even more." What has been will be again, what has been done will be done again; there* *is nothing new under the sun."* If there are no new babies, the entities must have been here from the beginning of time. This even goes to God Incarnate himself.

**"Melchizedek" and Jesus**-This Dead Sea Scroll helped us understand more about the ideas about reincarnation. It said, *"Melchizedek is reincarnated in the last days to destroy Belial (Satan) and lead the children of God to eternal forgiveness"*. The current New Testament books indicate that the one who destroys Satan is Jesus during his final return, so the Jews believed that Jesus had been reincarnated from the beginning and he was originally this Melchizedek guy.

# Christians

**Matthew Chapter 11 and 18-** *"For all of the prophets and the law have prophesized until John. And if you are willing to receive it, He [Jesus] is Elijah who was to come." -- 'Why then do the scribes say that Elijah must come first?' But he answered them and said, 'Elijah indeed is to come and will restore all things. But I say to you that Elijah has come already, and they did not know him, but did to him whatever they wished. So also shall the Son of Man suffer at their hand.' Then the disciples understood that he had spoken of John the Baptist."* This shows a strong belief in Reincarnation, so what happened to John the Baptist/Elijah?

**Matthew 17:1-13** *"After six days Jesus took with him Peter, James and John the brother of James, and led them up a high mountain by themselves. There he was transfigured before them. His face shone like the sun, and his clothes became as white as the light. Just then there appeared before them Moses <u>and Elijah</u>-* So, john, now Elijah came back to life, talking and completely visible, but then he vanish. These guys had been dead for hundreds of years. They had not been to heaven because Jesus told everyone that he was going to make heaven livable so you ask, "Where were they?" Well the next section tells us a little bit concerning Elijah and where he has been.

**Mark 10-**"*No one who has left home or brothers or sisters or mother or father or wife or children or land for me and the gospel will fail to receive a hundred times as much in this present age - homes, brothers, sisters, mothers, children and fields ... and in the age to come, eternal life.*" I know you are wondering how someone can get all new brothers, sisters, mothers, and children without getting a reincarnated body during the "present Age".

On and on I could go, but the Christian belief of Reincarnation shows pretty much what I have been talking about. People don't die as their consciousness is part of what we believe to be reality. It seems that sometimes we can even get a new body and live again. Everything was great until the Catholics came along.

**Catholic Rebellion-**Something happened when Constantine I started building a new type of Christianity. The Catholic Church outlawed and put to death those that preached reincarnation and quit emphasizing the only difference in Christianity call the Holy Spirit. If you died before getting this marvelous thing, they came up with something called purgatory, but they used it to get money rather than using it to help people understand the basic characteristics of death [Reincarnation, transmigration, the Holy Spirit, Heaven/Hell, and prenatal existence]. It wasn't long before much of the Christian beliefs were tainted. The fifth ecumenical council in 553 AD stated the following: If anyone asserts the fabulous pre-existence of souls, and shall assert the monstrous restoration which follows from it: let him be anathema. This anathema word means "something dedicated, especially dedicated to evil" so it was a bad word to use. Certainly, these saying had nothing to do with the New Testament teachings of reincarnation and transmigration which really are the reasons our reality stays constant.

Let me ask you this! Let's assume souls, like matter and energy, are not created or destroyed [except by the creator God]. If our reality is still the same, the souls are, generally, all here. Death is simply a released for the carnal portion to another reality that established reality according to anthropic science.

# After Death

There is an interesting verse in the Bible Romans 8:28- *God causes everything to work together for the good of those who love God.* This is saying hurricanes, Tornados, Volcanoes, and all the things that destroy thousands are for "their good". If we are to believe this, these souls are not lost, but just had a learning experience that might allow them to have a fuller "good" next life. When Judas Iscariot died, it was said that it would have been better for him to have never lived. Remember that Judas was one of the disciples of God Incarnate and he had a moment of weakness where he got money to show the Pharisees where Jesus was. He was so upset with what he did, he hanged himself. Peter was going around denying he even knew Jesus and the great Apostle Paul would soon begin killing Christians before he got blinded and had a change of heart. ON and on we can find many of the ancient Christians did far worse things than Judas so one would believe that the statement really is saying *"Judas would not have been so sad if he had been reincarnated as someone else during this time."* At one point the Babylonians, under King Nebuchadnezzar, were said to be "*doing God's work* "as they slaughtered many Jews in a power crazed effort to control the entire world. This also seems odd unless it is talking about reincarnation.

In the beginning, I had to establish the precedence of religions you understand so we can look at each of the elements with a level of authority. At the beginning of this book you may have thought I was Godless. Hopefully, you will change you mind as we go along as I am strongly religious and use the oneness of religion and science to help build a truer understanding of the mysteries of the Bible and science. The mystery of our reality is one of the hardest to understand while we are living it. Even after death, it probably is difficult to understand the nuance of life resonance to control elements of a perceived reality that changes ever so slightly as people go beyond the normal carnal living and change their resonance for a time.

**No Divine Insight?-**This detail is not coming from some flash of enlightenment that others claim to have concerning insights that no one else can fathom. I'm not saying God does not give divine wisdom if it is needed, I'm simply saying that most of the time, what people this is divine insight is nothing more than gas. I'm not trying to be coy or sacrilegious, but just look around. Hundreds of religious leaders study and seem to think they get heavenly guidance and contradict each other on every turn. God could "guide" only one of a thousand or so heartfelt opinions, presented on any subject. The other "opinions" of thousands of truly devout believers in God MUST BE wishful thinking. These devote people who truly want to help everyone they talk to, ARE NOT GETTING INSIGHT from God. Remember, God is the controller of hurricanes, floods, volcanoes, destruction, sickness, and death. At the same time he does everything for the common good. It sounds so strange when represented by most of the religious leaders of the day. We will examine how, why, when, and where God's control, love, and mercy MAKES sense. We not only can accept that there are miseries, but also that the miseries are TRULY for the GOOD. OK! I hope some of the information I will be trying to provide will have a hint of insight, but you will have to get the details ironed out on your own as there are some Preston-isms that simply can't be helped.

Here are my thoughts for whatever they are worth concerning death. People need to get the Holy Spirit "Light" or they cannot go beyond that Crown Chakra state that some of the Buddhists kind of people talk about AND they CANNOT enter the Universe of Heaven. That being said, the "common good" would be for as many people to get the Holy Spirit "Light" as possible and the only reason, I can think of for God waiting thousands of years to come back to the Earth to take his people home would be to get a higher percentage of people to go with him. Clearly, the percentage seems to be getting worse instead of better, so there is a strangeness to be reckoned. If the percentage is getting lower and lower, God would have returned sooner, so we can assume that the percentage is actually getting larger. That is where something we can call "re-entry of life" comes in. rather than death being simply DEATH.

# Conservation of Everything

This law will help us understand death for sure. In general, conservation means that, in an isolated system, a given physical quantity does not change with time. As a derivative, an especially important and useful conservation law is that matter and/or energy are neither created nor destroyed over time; they merely change form, and their sum total always remains the same. I now this seems to eliminate the possibility of ghosts needs to be addressed as it is confusing.

Ghosts or demons, have a life essence that cannot leave our universe. When a change occurred from this living entity to a dead one, something very strange had to occur to balance out the universe. When they died, they did not return to life. Oops! I'm sorry I said that before you were ready. Conservation Science tells us before one can have his consciousness and living body "Die" and equal measure of consciousness and life must be regenerated or he CANNOT DIE.

You might ask, "What about these angels that appeared out of nowhere? How does conservation of energy and matter account for them?" Many respectable people have seen angels from time to time who, apparently, leave their nonphysical realm to appear here in ours. If they can interact with our material environment, they must be at least partially composed of matter themselves since it is an observable fact that only matter produces the radiation, gravity, and mechanical forces that affects other matter. By disappearing in their "heaven world" and appearing in ours, many "scientists" indicate that they violate conservation of matter and energy in both worlds! "This simply cannot occur", the Scientist exclaims! For the strict energy conservationist, the whole "change from life to death or ANY human" gives him heartburn because energy seems to be lost so they came up with an interesting theory called Super symmetry.

**Super Symmetry**

If someone "dies in this universe, there must be a new life in the next universe to counter it. We will get into this next universe in a minute, but super symmetry as I explained in book 1 says that EVERYTHING is backwards in the adjacent universe. If something is made smaller here, that can be explained by having something get large over there. While death might be considered the lowest usable energy level or entropy of life in our universe, it is ATTACHED to the most vibrant life characterization in the adjacent universe. The Theory of Super Symmetry allows the transfer of reciprocal energy to another UNIVERSE. From it, life MUST go on or "LIFE Energy" would be destroyed when it can't be destroyed just like matter can't be destroyed. Your life energy simply cannot cease to exist. I know you are just saying---so what, another person simply is created and uses my life energy when I die.

That sort of is correct in that new people, personalities are created ALMOST every time someone dies. Fortunately or unfortunately there must be also a limbo place where consciousnesses react in the Heaven or our universe between states. Hopefully, I can explain the physics of all of this to you in an understandable way and also tell you about some very unique evidence that is being established by regression scientists. One in particular is a man named Michael Newton. I think you will love the details that he has uncovered. Souls seem to enter new bodies and sometimes we remember what happened before and "possibly" even what happened between lives.

# Soul Journey

Michael Newton is a hypnotherapist in California and he is good at it. His specialty is regression therapy. As many others have found, sometimes, fears or limitation we have in this life come from events that occurred in a previous life. Too many people have been regressed for it to not be considered a reasonable method of research and it cures people. Anyway!!! Michael started getting information from his patients that didn't make sense. They were talking about people they had contact with BETWEEN LIVES. Case study after case study, more and more data was collected and confirmed by the details presented by others. Soon, a reasonably clear picture of this purgatory type place you have heard about and dismissed. When events of a life are not completed in a reasonable way or the person feels anguish at something that happened or he did while alive, he relives those events on the other side to experience the other side and gain empathy. I'll bet you thought you got empathy because of your marvelous insight to your surroundings. Well!!! There is a growing amount of data that suggests that empathy comes from ACTUALLY living a life as the OTHER individuals. Thomas Maslow's Self Actualization level that forces the insight concerning those around us, may not be a carnally learned event, but is accomplished over many existences. Reincarnation taught in the Bible now makes more sense. Why would there be reincarnation unless there were new things that had to be thought to the reincarnated souls.

**Samuel** being summoned by the Witch of Endor from "purgatory makes sense and his being conscious makes more sense. Also, his ability to see the future makes more sense as we will investigate in a little bit.

**Reanimation-**The idea that when Jesus returns, the dead in Christ rise up out of their graves first as attested to in a large number of Old and New Testament scriptures, makes sense as there is somewhere to rise from that is not just dirt.

**Rebels-** This whole concept presented in the Bible and many other texts about the losers of the Heaven Wars being dead, but never being able to leave the Earth makes more sense.

I know this has been quite a bit to spring on you so let's start expanding on some of the more important themes before going into why some of this HAS to be or Physics would not work. Our first stop is re-entry of life.

# Re-Entry of Life

I know some call this reincarnation, but I wanted to be different. As I showed earlier, Jesus' Disciples believed in this characterization during the time they were being taught by Jesus himself, so one might believe that some form of reincarnation is possible. The strange discussions of the mysterious Melchezedek who somehow survived for thousands of years seem to indicate this characteristic and "John the Baptist" was thought to be reincarnation of one of the ancient prophets. Repeatedly, this seemingly absurd reincarnation keeps popping up in the Bible and many other ancient religious texts but most try to ignore it because it is uncomfortable. One might even wonder how the preflood people lived hundreds or thousands of years. If sequential life cycles were possible, the time constraints go away and the Bible begins to make more sense.

## God Waiting 2000 Years

If you have never wondered why Jesus is waiting 2000 years before coming back to earth, I think something is wrong with you. That is a long time. He told his disciple that he would be coming back VERY soon, so there is a mystery. My belief is simple. God wants as many to come with him as possible. He wants as many as possible to take something called the Holy Spirit or the LIGHT and be able to "cross over from hear to the place or universe called heaven.

The apostle Paul of the Bible indicated that --*"**Man thinks only of evil continually.**"* What he meant was that normal vibration level of people assure that Sex, Self centered, self awareness, and Survival are paramount rather than anything that is able to get you out of what one could call the three "Ss". Another way of saying this is that, most of the time, we are comforted by simple pleasures of carnal living. Unfortunately, I know that Paul was correct .If you die before you gain that "help" you are in trouble. You may get another chance, but at sometime, the chances are all gone.

## Consciousness Doesn't Die

Some may tell you that when you die, everything is gone, but just think about it. As I stated before Einstein theorized, along with many others, that things only happen if witnessed by a consciousness. Let's say that all the ancient consciousnesses died and we are living in a new world with brand new consciousness. Because we are all new to this game, our consciousness reality is now substantially different than the reality of a thousand years ago or a million year ago. If we assume that reincarnation is factual, every day our changing, new inexperienced consciousnesses drastically change reality as perceptions are altered from predecessor events locked

away in our consciousness, but still affecting reality. Many well tested theories all say the same thing.
- Conservation of energy and mass
- Conservation of consciousness
- Continual fight against the lowest states [ENTROPY] of all dimensional characteristics of the universe
- Knowledge that there is no difference between a dead DNA and a live one
- Loss of a segment of reality if a consciousness is lost

These all say the same things. They are well tested and they keep pushing us to *this* logical conclusion that allows physics to operate without getting torn down by the <u>absurdity of consciousness termination</u>. There can be no end to our consciousness.

Instead of thinking it reasonable to have the consciousness end as breath leaves a body, it becomes more and more reasonable to think that there is somewhat of a continuation of consciousnesses.

### What I Mean By This

This is certainly going to sound Biblical, but people aren't good at being the type of people God needs us to be and the type of people we, ourselves need to be. Christians believe that only this Holy Ghost can change us and because of our being bad, only God Incarnate's sacrifice could allow us to get this Holy Ghost. As a similar religion, Eckankar almost says the same thing except they start with a premise that people can get better without this Holy Ghost by learning from "teachers" that instruct us on how to improve ourselves. If one gets good enough, they can then, sort of get this Holy Ghost thing on their own. All along the way, we learn more and more about the nuances of living in a carnal and a non-carnal world.

*Many of the lessons we must learn require suffering.*

Hopefully, you can sense that God causes and allows suffering to HELP us learn, grow, understand, develop, and fulfill the requirements of God and to allow our reality to be fixed as the resonance of our various lives increases. Most religions seem to have this as a common thought. However, they typically identify the "helper" as not being of God, but as being others who have become enlightened. For this overview, either type of helper will demonstrate the same thing.

<u>God promotes good by allowing things we think of as evil exist.</u>

# Life and Death?

### Death

That was getting a little heavy so let's move away from here and just go into definitions. I think I had better figure out what death is in the general sense. Once we can establish that, our journey will become easy. Unfortunately, there seems to be a lot of different deaths and we may have to pick and choose.

**Hell is death-** *Not only is it defined as death, it is also determined to be everlasting death. I think I need to put that in a special place. While I will define it a little more than you are used to, I don't think I will get into it much as it is too scary for me.*

**Death transported to many hells and heavens-** *While most seem to be rejuvenated into new bodies after death, sometimes, before a return, investigations into a wide assortment of places has been possible. Enoch and many others left us with some of the details.*

**Recurring death and reincarnation-** *No! I'm not talking about coming back as a bug, however, I have known* *some people who came back as snakes. We will have to look at that type of death for sure.*

***Learning Transition-*** *On the same line as that above, many now have a belief that one must return to satisfy some mistake or yearning after death. I'll give you some of the details of this interesting thought process and some of the evidence.*

***Death and resurrection in the future-*** *This was promised by the Biblical accounts of Daniel, Thessalonians and Revelation. This one certainly is interesting and worthy of our thoughts.*

***Death is the halting of body operations-*** *While this one is one many people believe, the chances that we become nothing after heart stops looks less and less like death the more we investigate. Our investigation here will be limited to Near-death experience rather than the whole enchilada.*

**Daniel 11 and 12-***"Those who are wise will instruct many, though for a time they will fall by the sword or be burned or captured or plundered."* People will be teaching others even after they have fallen by the sword.

*"Some of the wise will stumble, so that they may be refined, purified and made spotless until the time of the end, for it will still come at the appointed time."*

EVEN the wise will "stumble" in one life, but they will have more chances until the end of time.

*"Multitudes who sleep in the dust of the earth will awake: some to everlasting life, others to shame and everlasting contempt."*

## Please read these three passage a number of times

The verses say, after many reincarnations, there will be a resurrection. If, after many stumbles, one finally takes the Holy Spirit, this resurrection will be neat and while you are "living", the resonance of life will be increased.

If this resurrection /reentry capability to increase the resonance of our reality was not so, the thousands of years God stays away does not make sense. Jesus continued to tell his disciples that the time of God's return was at hand. They thought it was to be in their lifetime. In one way, it was as more and more believers are collected for his final return and we are not aware of any time spent between lives or in previous lives, so to us the time is still at hand.

The Holy Spirit described in the Bible and this thing called Light by many ancient documents and tales seem to be the same thing and it is the most vital thing we must capture to understand and live peacefully in whatever type of after life there can be. To be able to receive this "light" you must increase your resonance.

### Life

Certainly, there are many more commonly believed and exotic life and death ideas, but I really want to focus on increasing your vibration level and developing sustainment or resonance to allow you to be aware of a more true reality where people can move mountains and walk across water. Let's begin with a review.

"The Meaning of Life", using a large number of hypnotic regression studies show strong evidence of this reentry or reincarnation into the carnal world after "death" where some type of out of body teaching and comforting allow re-entry into life sort of as a second chance. Don't get me wrong here.

Anyone thinking he or she will get more time to live as debased as one can, will assuredly lower the resonance of our entire reality. Then, will find that the carnal world is continually getting farther and farther away from the spirit universe. This Holy Spirit has a more difficult time advising and "giving the light" as no one is listening to the consequences of life without God, without light, and without hope after death. We experience life as a dissipating element. When we get to its end, it somehow disappears.

# Dissipating Life

Einstein indicated that our very existence was determined by consciousness. As an example let me say, "If no one was in the forest to hear a tree falling, there was NO tree falling." Added to this now common belief, there is the theory that if light and life only went forward in time, soon, time and life would travel beyond the realm of our universe and we would lose both forever. The universe would get dimmer and less and less "life" could be generated. Some tie this theory to the Law of Entropy which forces EVERYTHING into its most disorganized state.

<u>Life's most disorganized state could be considered death.</u>

His idea of the universe had an ever expanding dynamic. Everything would simply expand to the edge of the universe and be lost. This included thing that held mass together, the thing that allowed for Light, and finally the characterization of life. In the end, nothing would be left.

After a time, there would be more death than life and the universe would be overtaken by death. As death becomes the dominant characteristic of "reality, it would be pushed out to the limits of the NEW universe until nothing is left and its opposite become the ENTROPY of the New Age.

## *Life would be the most disorganized death there could be.*

If you are trying to put down the book I would not blame, you, but I have to get you thinking weirdly or discussions of death will not be healthy. Speaking of that, why did you start reading this book in the first place? Hopefully, you can see there is a fine line between life and death. The Bible informs us, great thinkers like Einstein help confirm this, the Law of Entropy might not turn us into tiny thought molecules after we die.

By the way, the universe is not getting dimmer as the Law of Entropy and the theory of expanding universal elements would have one to believe, so something else is going on. In the last couple of books in the "Vibrational Universe series", I talked about Milo Wolff and his more comfortable descriptions of an adjacent universe [Heaven]. Let me just bring out a couple of his discoveries. As a physicist, he described how there are balancing in-waves [sort of inverted matter or matter going backward in time] from this linked universe that "refill" our universe as "Light" leaves it at the universe limits. We must have this "joined universe" to stabilize the loss of forward time matter, forward time energy, and forward time life. This is what stabilized ENTROPY. We might imagine that when death leaves this universe it must remerge as life in our adjacent universe or there could be no stability. Life and death act the same, because they ARE the same.

Let me say it another way. Life and death act similar to "positive timed light" leaving our universe only to be regenerated by sort of a "negative timed light" we call "fermio-nucleaic force" and it acts the same with particles leaving out universe and the opposite we call electro-magnetic force enters. I know that's confusing, but just know that we enjoy a stable universe because it is rejuvenative. We also enjoy life that is rejuvenative in the same way. As life finally leaves our universe, there MUST BE a balancing life force regenerated from another place.

**IT MUST BE THAT WAY *or the universe would not continue*.**

While this must be true, how long our conscious minds remain in this universe is a mystery. If we are to believe the ancient Biblical teachings, one might believe all consciousnesses have stuck around for thousands of years waiting for this final exchange. While they, apparently can sometimes experience existence without life, generally speaking, on sort of "sleeps" until reentry into another "Life" or exchange to our adjacent universe "Heaven".

# How Many Heavens?

As I stated, a number of texts talk about DEATH as traveling from hell to heaven and back again, and there seems to be some consistency about there being a large number of both universes. Some may not realize that the King James Version of the Bible referenced at least 5 heavens, while other Jewish references detail as many as 10 heavens and the Biblical writings certainly do not preclude the possibility of more than 5 heavens while Mormons describe only 3.

Here is an interesting note. In the Jewish texts, not every "level" of heaven to the early Jew was a place of comfort as one may expect from today's preconceived notion of heaven.

Just about all regions of the world talk about many heavens. It was as if the universe "heaven" was filled with many "planets" to live on or something similar to that. I'm not getting into Heaven in this book as it simply confuses us when we are thinking about a very confusing topic in the first place. Let's just know that scientifically, there must be a "linked" universe to us and keep it at that as we investigate different Death models.

# Dreaming

Today's models of the universe are characterized by 3 independent dimensional blocks. One creates things. Another creates forces that allow interaction of things, chemical attractions, magnetics, electricity, photons, Radio waves, etc. All that is interesting, but now we know that there is still a third three-dimensional characterization required to build a universe. This last component allows things to live, think, dream, die, and, most importantly allow for a carnal existence---outside what we call reality.

Go to sleep! Can you still see things? Where does the light come from? Where does the entire world come from? I know you are just saying imagination, but there is much more to it. There is actually a change in the world from you dreams. The changes are miniscule, but one could sense them if they had some type of meter to test that sort of thing. The light you see in your dream is just as real as the light you "sort of perceive" when your eyes receive electromagnetic waves that reflect off surfaces. That jumbled mess of vibrations going forward and backward and side to side and being absorbed and changing wavelengths is a poor excuse for something we call light. What you really are interested in is perception. Perception of a dream and perception of what you have defined for yourself from all the electromagnetic waves are still just perception.

Here is a question. What color is red? The answer is way more complex that you would imagine someone can take photons that are vibrating a 700 nanometers per cycle and define red, but your conscious mind puts the dazzle to those vibrations, not the photons. My red probably is way different that red to you, but life is similar for all of us who have a similar resonance and this common similarity is continuous whether we are considered alive or dead.

# Why Continuous Life

Let me give you a different way of looking at life. Assume that accelerating matter in space creates something we could call an 'acceleration field'. This is a calculation analogous to the 'electric acceleration field' that produces a force on an accelerated electric charge --- so we can assume that the 'matter acceleration field' in space produces a force on an accelerated mass. In this situation, energy is transferred between the mass "fields" and electrical "fields". The force of the electric field and the force associated with the accelerated mass is an equal and opposite interaction.

*We continuously have an equality of force, mass, dimension and time.*

Additionally, we MUST have an equality of consciousness and life.

*Some of the characteristics are going forward in time and an <u>equal amount</u> are better characterized as going backward in time.*

The <u>opposing interactions are why our universe stays together, why light stays light, why time never is destroyed AND why our consciousnesses cannot leave the universe</u> or the perfect condition needed to support the universe will be lost and the universe will cease.

Because of this factor, we can define things in many ways. We can either say that the acceleration changes the frequency of the matter waves by the Doppler effect "or" time is shifted to produce this Doppler affect. Both are the same thing. So here Quantum Mechanics was born.

### Quantum Mechanized Life

If Life has control over existence as has been theorized by just about everyone to date. This book is specifically about that portion of our Living. As were increase our resonance, everyone benefits. Let's put it this way if you are around a "positive thinker" some of that will rub off on you.

### New Car

If you want a new car and can visualize the new car--- YOU WILL GET IT! It is "doubting the outcome" or tying you needs to simple desire that limits your capability. People who are with you when you are in this higher state of resonance will benefit as reality is "modified" near where you are.

## Around Jesus

The biggest resonance increaser of them all was God Incarnate. While he had to be 100% man, he also could affect reality as the increase the frequency of his life resonance. Those near him felt it, lived it, appreciated it, and gained insight about reality as they stayed near him as he modified reality to make wine, heal sick, walk on water, and bring people back to life.

# Life in Death

Where does existence come from when we die? Instantly, <u>unconstrained by time space</u>, life must be reformed if it ends. <u>If death is the absence of life, there could be no death as life cannot be reduced</u>. This does not mean that all life has a physical form, but it does mean that no conscious soul can leave this place until there is a controller that reconstructs the universe. This would be the Resurrection time controlled by God Incarnate. During this resurrection, those who had not accepted the "Light" cannot leave. Because the universe is reconstructed at that time, those who are not in the light and having a resonance that can be in a new reality are now in trouble as they will no longer affect the universe. Completely alone, separated from natural elements of existence, these consciousnesses are in something we can call HELL.

I'm not talking about a lake of fire that you are told about in church. I'm talking about something worse than we can imagine. While many don't like to accept it, we can only modify the "feeling" of our reality. Our creator God holds our universe together. Separation from him means the universe structure comes apart for those unfortunates. Our reality would blur. Interaction with other conscious entities would become strained, and finally, each consciousness would be alone, desperately trying to link up to others in this same existence. Without universal structure, simply holding one's self together as an entity would be a full time, unimaginably difficult job. Don't worry about fire in a Godless universe. Fire could not exist.

## Pigs

Let me give you an example. On a number of occasions, Jesus met individuals that had been invaded by the consciousnesses of a group of people originally called the Anak. The Anak enjoyed life for a time, but because they had participated in a war against Heaven, when they died, they became something we call demons. After death they could not experience "existence" at all unless they could enter a living entity and share the experience. Jesus would find these guys and force them back into what must have been like hell. One group of a thousand of these demons begged to be sent into pigs rather than emptiness as the loss of "reality" completely was too horrible to sense. If it is so bad thousands want to experience existence inside a pig, it was bad, but now we are talking about agony over thousands of years.

***Think of it this way, Hell is when you desperately need to find EVEN a pig to live in and there is no such thing.***

# God and the Trinity

I know some of this book sounds religious, but it is not meant to be a religious book for conversion and all of that; however, I would like you to be cautious about experimental conscious level transfer. Without a guide, changing your resonance level can be dangerous. Remember that these demons need company and they will do anything to have it. Some are considered "Spirit Teachers" of the Eckankar and similar religions. Being able to free your 'Soul" as your consciousness vibrates fast enough can be problematic as some talk about some visitors that are with you when you return. Most of the time these guys don't bother you as they don't want someone to remove them, but I am told they are still very disconcerting if they do share your "self". Yes; reality can be modified, but sometimes the outcome is not exactly what we want or need. That is where that "light" comes in to guide you as you become enlightened. Think of a person as being a trinity of entities; a soul, a "carnal self", and this spirit or light. Discarding the control of "self" is a good thing. Having help with the "spirit" can make it much better. Speaking of trinity, have you every wondered when God indicated he made man in his image as a trinity? The "real creator" God is the same as the real person [soul], Jesus incarnate is the same part as our Self, and the holy spirit and our spirit share similarity.

## God must be three entities

Just think about yourself in a 10-dimensional universe. You have an ethereal self, a force self and a particle self just like everything in the universe. God made it that way. Each of these "LIFE dimensional dynamos" is mutually perpendicular so we only comprehend one self. The Particle self is static and requires a force to be recognized [we can call that part the "Self", but if one looks at the kinetic component of existence or "force" part existence drive all things that are accomplished in our universe [this would be the light or spirit]. That brings us to the ethereal of life [the soul the essence of life]. Every particle; every force, and time itself all must be made as a trinity. God <u>Must</u> be a trinity.

### Think of God in the same way.

*The Bible calls Jesus "the word of God". His WORD or action would be the touch component or Particle Self.*

*The Holy Spirit as the LIGHT of God. The light portion, not to be confused with electro-magnetic vibrations we collect as we generate light, of Holy Spirit could be considered the Force Self.*

*The head of everything CREATOR GOD could be recognized as life itself. – He is the Ethereal Self or the soul of our Creator.*

Just like us, the Creator God modifies and creates reality and our little souls change it very slightly. No matter how you slice it, trying to make God out to be less than a trinity is impossible. While there are some similarities to the trinity that is man, there is no great similarity. That being said, we are made in his image—the image of a trinity.

As Anthropic Science tells us, without life, nothing exists. Do away with God and there is complete nothingness.

# Final End New Beginning

Of course, like everyone else, I don't completely know what happens after death, but there is a peculiar thing we recognize from Biblical texts that sometime we ignore.

In the book of Revelation, it states that *"when God returns to the Earth in glory, the dead in Christ will be raised"*. After "sleeping" for thousands of years, these people wake up and rise up into the air so there is a mystery here and it seems to go along with my previous statements.

**What happens to people who are to be raised up?**

Some tell us that they are in a state of slumber with **halted time.** Several scriptures say *"it is like being absent from the body and present with the Lord"*---so time doesn't pass when you are not in a body.

Here are my thoughts for whatever they are worth. People need to get the Holy Spirit "Light" or they cannot go beyond that Crown Chakra state and enter the "Spirit Universe of Heaven". Therefore, if the Biblical teachings tell us anything it is that God does everything for "good" so he has been waiting around for thousands of years helping and hoping for a higher percentage of souls to get joined with his "light/spirit".

If the percentage of "accepted ones" is getting higher, only the ones that are either not giving in or are given another chance to learn was they need remain on this less than holy Reality.

Don't be afraid of life or death. Take control of both of them. Make your life and your death count. Work to increase your vibrational resonance and become more in-tune with our creator. Separate from Carnal living and walk on some water.

# Conclusions

Hopefully, the main conclusion you have gotten from this book is that life is far greater than you previously believed. It's not bottled up inside some DNA. If you want to control your destiny, you can. To do it you must trust in something besides yourself and you must actually try to live as part of the universe rather than as a single individual trying to survive. Other import aspects that were brought out include the following.

**BORONS**-Early attempts at defining the unified particle always ended in failure. Bayrons [electrons] were broken down into Borons [the smallest particle containing all elements of existence].

**FERMIONS**-Unfortunately, the Borons could be broken down into FERMIONS which had component parts of particles but not everything. The one usually used as an example is the graviton which has gravity, but no mass to support the gravity.

**Not Fermions**-Unfortunately, there were many types of fermions so something else made them.

*All matter is made up of vibrational fields instead of particles.*

Vibration levels determine the character of a particle. As a photon vibrates faster, its characteristics chance from visible light all the way to massively dangerous COSMIC rays that can go right through your body.

**Frequency Effect**-If the electro-magnetic vibrational pattern is excited faster, the photon becomes a particle. If a particle vibration is increased it become a wave.

**Fast Vibration**-As the vibration still increases in speed, the particles get larger and larger. Gold, for instance, is vibrating much faster than helium.

**Inverted vibrations**-Just like noise reducing headphones make sounds disappear, vibrating particles become invisible to each other as they become in 180 degrees out of phase.

**Parts of the Universe**- Everything in the universe can be broken down into vibrations. This includes not only the things we can see like planets and tiny little pebbles, but also electricity, magnetism, Photons, Nucleons, the spirit, the soul and even life.

**Christian Religion**-You should have also gotten a better understanding about how the Christian religion fits with all of this stuff. How God incarnate came to reestablish the link between Heaven and Earth. How demons could be understood. How reincarnation could be addressed. How the final rapture of the Church or the 2nd coming of God incarnate could fit into a world not of 4-dimensions as you once thought but in a world of 10-dimensions.

**Not Religion**-This was not a religion book, but religion must coincide with our mathematical models or we have one of 2 problems. Either the science is wrong or the religion is wrong or incomplete.

**Soul Vibration**-Beyond that, everything in the universe can be broken down into vibrations. This includes not only the things we can see like planets and tiny little pebbles, but also electricity, magnetism, Photons, Nucleons, the spirit, the soul and even life.

**Light is not Light**-We looked at how light was different than what has been imagined as light in the past. While the apparent affect of light can be initiated by placing stress on particles with electromagnetic [In-waves], the electromagnetic waves do not cause light. They cause electro-magnetic fields. Light is established in our consciousness. To that end, the color red is completely different to all people. While we define it as the same, what might be a more vibrant red hue might be interpreted as less bright by another or even a completely different color by an animal seeing the same vibrations.

**Electro-magnetic vibrations**- don't light up anything. While they generally are present during feelings of light and visual comfort, sometimes, light can be sensed with our eyes closed or during near death experiences showing eyes are not the important part.

**Sideways Light**- We discussed how light is transposed from normal life and if time were viewed sideways, the thing we call light would appear as a solid mass while life would simply have excursions where an entire lifetime could be viewed simultaneously.

**Life Not DNA-** As we started studying life, it became apparent that life was not the same thing as DNA. Dead DNA and live DNA are similar in structure and forces on one verses the other would cause similar reaction.

Life was a different dimensional dynamo from matter and forces on that matter. While it is different, nothing can be established in our universe unless life is instituted and a general acceptance by the combination of the conscious group must be assured before a true world can be generated. As people try to veer away from the common knowledge or viewpoint our world changes. When God stated that "faith of a grain of mustard-seed would move mountains" tells us how important our conscious minds are to our reality.

**The Anthropic Universe Theory-** Tells us that each of us hold the world together. We cannot die or a piece of the universe will be lacking. Conservation of Energy won't allow it.

**Life Between Lives-** The old Purgatory looks a little different than described in many Church Dogma. Instead, substantial research has shown that people must learn more than they typically learn about the non-carnal existence so they reenter life several times.

**God Exists-** Unfortunately, the universe cannot exist alone. A controller keeps everything from slipping back into entropy and disappearing. This God is the creator of everything, I hope that you understood how very important it is for the God stability to work in this universe. In-waves and out-waves stabilize the nonliving. God stabilizes the living/conscious elements of the universe.

**God is Light-** In the form of the Holy Spirit or Holy Ghost; we found confirmation that this type of light was extremely important in not going to what we call Hell.

**Death Isn't Death-** We finally looked at this thing we call death. We, possibly, can use death as an opportunity. There seems to be some? Chances to gain the "light" from the Holy Ghost that allows us to be transferred to another universe [Heaven], but no one knows how many chances and it seems that it is getting harder to become en-"Light"-ened by the Holy Ghost. If one gets the right credentials, one can be transferred to a better place.

**Hell is Worse Than Bad-** We found that this whole lake of fire thing is not nearly as bad as not being part of existence that seems to be a truer definition of Hell.

**Resurrection-** People who finally understand the "light" don't have to experience the nightmare of no stable universe and no interaction between entities as we can imagine this Hell to be.

I hope this book has been useful to you when trying to interpret what otherwise was considered anomalous in our world. If too many things don't fit, you must try to change your definitions. This is a new definition.

With that, I must end this book.

# About The Author

Steve Preston is a long time author of scientific, esoteric facts. His series on the creation of mankind is shown below. The series focuses on the painful truths rather than whitewashed details that make us comfortable. If you are interested in the truth instead of comfort, please continue to read and, while you are at it, review other works by Mr. Preston as shown below.

**Four Part Series "Vibrational Matter"**
Vibrational Matter
Our 10-Dimensional Universe
Walk Though a Wall and Time
The Meaning of Light and Life
**Eight Part Series "History of Mankind"**
The First Creation of Man
The Second Creation of Man
The Creation Of Adam And Eve
The Antediluvian War Years
Man After the Flood
Life After the Babel War
A New View Of Modern History
The 20th Century To The End Of Time
**Truth Series**
The Truth About Dinosaurs

The Truth About The Earth
The 7 Destructions of the Earth
The Truth About the Heaven War
Truth About Dinosaurs
Who Really Discovered the Americas?
God Didn't Make The Ape

**Planet Series**
When Did People Live on The Moon?
Evolution of the Planets
The Day Venus Exploded
Living on Mars

**Odd Series**
The Book Of Odd
More Oddness
Why Are There So Many Anomalies?
Stupid Science

**Other Works**
A Closer Look At Genesis
A Closer Look At Lincoln
Adam, Lilith, and Eve
America's Civil War Lie
Ancient History of Flying
Behind the Tower of Babel
The Funny Book of Law
When Giants Ruled the Earth
Lizard People
Live and Die the Right Way
Death Without Death

www.ingramcontent.com/pod-product-compliance
Lightning Source LLC
Chambersburg PA
CBHW051646170526
45167CB00001B/356